ExoticsCon 2022

Building Exotics Excellence. One City. One Conference.

AA005351

Denver, Colorado, USA
14-18 August 2022

ISBN: 978-1-7138-7802-5

Printed from e-media with permission by:

Curran Associates, Inc.
57 Morehouse Lane
Red Hook, NY 12571

Some format issues inherent in the e-media version may also appear in this print version.

Printed with permission by Curran Associates, Inc. (2024)

For permission requests, please contact Association of Reptilian & Amphibian Veterinarians
at the address below.

Association of Reptilian & Amphibian Veterinarians
5003 SW 41st Blvd
Gainesville FL 32608-4930
USA

Phone: 480-703-4941

info@arav.org

Additional copies of this publication are available from:

Curran Associates, Inc.
57 Morehouse Lane
Red Hook, NY 12571 USA
Phone: 845-758-0400
Fax: 845-758-2633
Email: curran@proceedings.com
Web: www.proceedings.com

TABLE OF CONTENTS

NUTRITION AND REPRODUCTION

ARAV MASTERCLASSES

ANESTHESIA, SURGERY, AND IMAGING

CLINICAL PATHOLOGY

POTPOURRI

HYBRID: MIXED TOPICS

HYBRID: EMERGENCY AND CRITICAL CARE

TECHNICAL TOPICS

Gap in pagination due to formatting issues.

Pages 1-325

Section 26

ARAV
Plenary Session

Paul Gibbons, DVM, DABVP (Reptile and Amphibian Practice)
Session Chair

Intra-Articular Conversion Electron Therapy in the Management of Osteoarthritis in a Komodo Dragon (*Varanus komodoensis*)

Rob L. Coke, DVM, Dipl ACZM, DABVP (Reptile and Amphibian Practice), CVA, and Bekky Muscher-Hodges

Session #2642

Affiliation: From San Antonio Zoo, 3903 N. St. Mary's Street, San Antonio, TX 78212, USA.

Note: Abstract format selected by author(s). See Front Matter for description of conference proceedings manuscripts submitted as scientific abstracts.

Abstract: A 28-year-old, male Komodo dragon (*Varanus komodoensis*) presented with a history of decreased range of motion of both stifle and cubital joints with palpable crepitus. Radiographs revealed moderate bilateral stifle and cubital osteoarthritis with generalized periarticular osteophytes. For the previous 8 years, acupuncture and rehabilitation techniques have been utilized to manage the chronic mobility issues, but over time limited improvement was noted. Quality of life was evaluated, and a novel therapeutic modality was elected to be performed. Both stifle and cubital joints were injected with homogenous Tin (117mSn) stannic colloid in ammonium salt (Synovectin OA®, Exubrion Therapeutics, Buford, GA 30518). This compound is a mild radioactive isotope that is classified as an intra-articular conversion electron therapy. It has been approved for use in canine cubital joints for arthritic conditions and can only be administered by veterinary specialists with a nuclear medicine permit. After injections were administered, the Komodo dragon was placed on temporary quarantine with limited staff interaction due to the radioactive isotope. Eight months after injections, the Komodo dragon exhibited markedly increased activity and mobility. Currently, the Komodo dragon continues to show improvement in mobility and attitude. This case is the first known documentation of this compound being clinically used in any exotic animal.

Clinicopathologic Features of Cutaneous and Subcutaneous Sarcomas in Captive Snakes

Michael M. Garner, DVM, DACVP, and
Juan Francisco Muñoz, DVM, PhD, DACVP

Session #2556

Affiliation: From Northwest ZooPath, 654 West Main Street, Monroe, WA 98272, USA (Garner), College of Veterinary Medicine and Biomedical Sciences, Colorado State University, 2450 Gillette Drive, Fort Collins, CO 80526, USA (Muñoz).

Note: Abstract format selected by author(s). See Front Matter for description of conference proceedings manuscripts submitted as scientific abstracts.

Abstract: Cutaneous soft tissue sarcomas (STS) and histiocytic sarcomas (HS) are relatively common in domestic species but have been uncommonly reported in snakes.[1] In this retrospective study, we describe the clinicopathologic features of these neoplasms in captive snakes. Forty-two STS were selected from the Northwest ZooPath archives. The age of 34 snakes ranged from 1 to 35 years. Colubrid (19) snakes were overrepresented. Fibrosarcoma (26) was the most common histologic subtype. These neoplasms were graded using the French grading system, which has been validated in canine STS and is based on three parameters: tumor differentiation, mitotic count, and tumor necrosis. Grade II (27 cases) were more common than grade I (5 cases) and grade III (10 cases) neoplasms. Euthanasia or death was attributed to STS in 15 snakes with grade II neoplasms (35.7%) and 4 snakes with grade III neoplasms (9.5%). Metastasis was observed in 6 snakes with grade II STS (14.2%) and 3 with grade III STS (7.1%). Two STS developed at microchip implant sites. Fourteen sarcomas with giant cells were assessed by immunohistochemistry for CD204, Iba1, actin, and desmin. Three of these neoplasms were further characterized as histiocytic sarcoma (CD204+ and/or Iba1) and the remaining neoplasms were interpreted as muscle origin (actin+ and desmin +) and fibroblastic/myofibroblastic origin (actin+ and desmin-). Histiocytic sarcoma was observed in elapid (2) and viper (1) snakes. Our findings suggest that STS are likely common in snakes and grade II or grade III neoplasms are more likely to metastasize and/or contribute to clinical decline and death. Histiocytic sarcomas are rare.

References

1. Dietz J , Heckers KO , Aupperle H, Pees M. · Cutaneous and subcutaneous soft tissue tumours in snakes: a retrospective study of 33 cases. *J Comp Pathol.* 2016;155(1):76-87.

Section 27

Infectious Disease

Sue Horton, DVM;
Shane Simpson, BVSc(Hons), GCM(VP)
Session Chairs

Rhodotorula spp. Yeast Mycoses in Reptiles

Lisa Harrenstien, DVM, Dipl ACZM,
James F. X. Wellehan, Jr., DVM, MS, PhD, Dipl ACZM, Dipl ACVM (Virology), Dipl ACVM (Bacteriology and Mycology),
Michael Garner, DVM, DACVP,
James E. Bogan, Jr., DVM, DABVP (Canine and Feline Practice), DABVP (Reptile and Amphibian Practice), CertAqV,
Jean A. Paré, DMV, DVSc, Dipl ACZM, DipECZM (Zoo Health Management)

Session #2505

Affiliation: From Avian and Exotic Veterinary Care, Portland, OR 97213, USA (Harrenstien), Zoological Medicine Service, University of Florida College of Veterinary Medicine, Gainesville, FL 32608, USA (Wellehan), Northwest ZooPath, 654 W. Main St., Monroe, WA 98272, USA (Garner), Central Florida Zoo & Botanical Gardens, Sanford, FL 32771, USA (Bogan), Wildlife Conservation Society, Bronx, NY 10460, USA (Pare).

Note: Abstract format selected by author(s). See Front Matter for description of conference proceedings manuscripts submitted as scientific abstracts.

Abstract: Rhodotorula species are yeasts that are found naturally in a wide variety of habitats, including soil, air, plants, and aquatic ecosystems including hypersaline lakes.[1] *Rhodotorula* species have historically been considered to be nonpathogenic, found in and on nails, skin, oropharynx, urine, and feces of asymptomatic human and nonhuman species (rhea, cockatiel, chicken, sheep, cattle, southern sea lion, domestic cat, domestic dog);[2-4] however, in the last 35 years *Rhodotorula* have emerged as opportunistic pathogens, found in immunocompromised individuals, those with indwelling plastic implants (e.g., central venous catheters), and also in immunocompetent individuals.[1,4] Correct speciation of *Rhodotorula* in a clinical setting, and subsequent selection of best treatments, have been hampered by the relatively low specificity of traditional laboratory techniques. Speciation accuracy is much improved with the use of matrix-assisted laser desorption/ionization time-of-flight mass spectrometry (MALDI-TOF) and molecular methods (PCR and DNA sequencing).[5,6] Using these methods, *R. mucilaginosa* is the most commonly identified of the genus.[4,7] *Rhodotorula* typically shows biofilm formation, resistance to the chemical effects of bile, and multidrug resistance (including resistance to itraconazole, fluconazole, voriconazole and caspofungin), with amphotericin B being the drug of choice for treatment.[3,5,7,8]

Here we describe a fatal case of *R. mucilaginosa* fungal choledochitis in an adult bearded dragon, as well as incidental dermatitis in another adult bearded dragon. For context, we also describe *Rhodotorula* dermatitis in an emerald tree boa. In addition to this case series, we also present a review of relevant literature regarding diagnosis and management of *Rhodotorula* mycoses in reptiles.

References

1. Hof H. Rhodotorula spp. in the gut-foe or friend? *GMS Infect Dis.* 2019;7:Doc02. 10.3205/id000042.

2. Cordeiro RA, de Alencar LP, Brilhante RSN, Antifungal susceptibility of emerging opportunistic yeasts and yeast-like fungi from *Rhea americana. Can J Microbiol.* 2019;59:577-580.

3. da Cunha MML, dos Santos LPB, Dornelas-Ribeiro M, Vermelho AB. Identification, antifungal susceptibility and scanning electron microscopy of a keratinolytic strain of *Rhodotorula mucilaginosa*: a primary causative agent of onychomycosis. *FEMS Immunol Med Microbiol.* 2009;55:396-403.

4. Wirth F, Goldani LZ. Epidemiology of *Rhodotorula*: an emerging pathogen. *Interdiscip Perspect Infect Dis.* 2012;465717. doi:10:1155/2012/465717.

5. Nunes JM, Bizerra FC, Carmona e Ferreira R. Molecular identification, antifungal susceptibility profile, and biofilm formation of clinical and environmental *Rhodotorula* species isolates. *Antimicrob Agents Chemother.* 2013;57(1):382-389.

6. Qian J, Cutler JE, Cole RB. MALDI-TOF mass signatures for differentiation of yeast species, strain grouping and monitoring of morphogenesis markers. *Anal Bioanal Chem.* 2008;392(3):439-449.

7. Gharaghani M, Taghipour S, Mahmoudabadi AZ. Molecular identification, biofilm formation and antifungal susceptibility of *Rhodotorula* spp. *Mol Biol Rep.* 2020;47:8903-8909.

8. Krzysciak P, Macura AB. Drug susceptibility of 64 strains of *Rhodotorula* sp. *Wiad Parazytol.* 2010;56(2): 167-170.

Diagnosis of Yellow Fungus Disease in Bearded Dragons (*Pogona vitticeps*) Using Next-Generation DNA Sequencing

Gina Vecere, DVM,
Shahar Malka, DVM, DABVP (Avian Practice),
Nitzan Hakim, DVM,
Nicolle Lofgren, VMD,
Janina A. Krumbeck, PhD

Session #2547

Affiliation: From the Long Island Bird and Exotics Veterinary Clinic, 333 Great Neck Road, Great Neck, NY 11021, USA (Vecere, Malka, Hakim, Lofgren), MiDOG LLC, 17062 Murphy Avenue, Irvine, CA 92614, USA (Krumbeck).

Note: Abstract format selected by author(s). See Front Matter for description of conference proceedings manuscripts submitted as scientific abstracts.

Abstract: Yellow fungus disease is a common dermatologic condition in reptiles, most often diagnosed in bearded dragons (*Pogona vitticeps*). The disease is often called CANV, as it was traditionally thought to be caused by the fungal organism *Chrysosporium* anamorph of *Nannizziopsis vriesii*;[1] however in recent years numerous other organisms have been identified that can cause lesions classified as yellow fungus disease, the most common being *Nannizziopsis guarroi*.[2,3] Bearded dragons infected with yellow fungus disease often present with raised, crusted, yellow to brown dermatologic lesions which can flake off leaving ulcerative lesions.[1] Traditionally polymerase chain reaction (PCR) has been used to diagnose CANV in bearded dragons; however, PCR results can be misleading in cases of yellow fungus disease due to the large number of fungal organisms that can cause clinical illness. In this case series, next-generation DNA sequencing was used in conjunction with a PCR for CANV to diagnose yellow fungus disease in five individual captive bearded dragons. The Illumina MiSeq (MiDOG LLC, Irvine, CA, USA) microbiome DNA sequencing technology that was used provided a microbiome panel listing all bacteria and fungi in the included skin samples. In all cases, fungal organisms within the *Nannizziopsis* genus were identified via DNA sequencing and were negative by PCR. We postulate that DNA sequencing is a more sensitive diagnostic modality than PCR for the diagnosis of yellow fungus disease in bearded dragons, especially given the wide range of potential causative organisms for this disease.

References

1. Mader DR, Divers SJ. *Current Therapy in Reptile Medicine and Surgery*. St. Louis, MO: Elsevier Saunders; 2014.

2. Gentry SL, Lorch JM, Lankton JS, et al. Koch's postulates: confirming *Nannizziopsis guarroi* as the cause of yellow fungal disease in *Pogona vitticeps*. *Mycologia*. 2021;113(6):1253-1263.

3. Schneider J, Heydel T, Klasen L, et al. Characterization of *Nannizziopsis guarroi* with genomic and proteomic analysis in three lizard species. *Med Mycol*. 2018;56(5):610-620.

Evaluating the Sensitivity and Specificity of *Cryptosporidium serpentis* Probe-Hybridization Quantitative PCR

James E. Bogan, Jr., DVM, MVS (Conservation Medicine), DABVP (Canine and Feline Practice), DABVP (Reptile and Amphibian Practice), CertAqV,
James F. X. Wellehan, DVM, PhD, Dipl ACZM, Dipl ACVM

Session #2592

Affiliation: From Central Florida Zoo & Botanical Gardens, 3755 W Seminole Blvd, Sanford, FL 32771, USA (Bogan), Department of Comparative, Diagnostic and Population Medicine, College of Veterinary Medicine, University of Florida, Gainesville, FL 32608, USA (Wellehan).

Note: Abstract format selected by author(s). See Front Matter for description of conference proceedings manuscripts submitted as scientific abstracts.

Abstract: The ante-mortem diagnosis of ophidian gastric cryptosporidiosis can be elusive if the proper diagnostic test is not performed on the proper sample. Molecular diagnostics (e.g., PCR) have been shown to be superior to other diagnostic tests (e.g., acid-fast stain, IFA, and ELISA) in detecting the presence of *Cryptosporidium* in a sample. Samples collected from the stomach three days after a meal have been shown to have a higher likelihood of detecting the presence of *Cryptosporidium* than other samples (e.g., cloacal swab). A probe-hybridization quantitative polymerase chain reaction assay specific for *C. serpentis* (qPCR) has been developed and shown to be extremely sensitive, but clinical sensitivity and specificity for this test is lacking. To approximate the sensitivity and specificity of the *C. serpentis* qPCR, the medical records from the Orianne Center for Indigo Conservation were reviewed. Between November 2015 and June 2021, 63 eastern indigo snakes (*Drymarchon couperi*) were necropsied. Of these 63 snakes, 11 had qPCR performed on gastric biopsies collected at the time of necropsy, 8 had qPCR performed on samples collected by gastric swab within 35 days prior to necropsy, and 44 had qPCR performed on samples collected by cloacal swab within 100 days prior to necropsy. The qPCR results were then compared to the post-mortem histological findings, where all three sampling techniques had a 100% specificity. The sensitivity was highest in samples collected at necropsy (100%, CI: 63.06-100%) followed by the ante-mortem testing: gastric swab (87.50%, CI: 42.13-99.64%) and cloacal swab (65.38%, CI: 44.33-82.79%).

Differences in Herpesvirus Detection Rates in Pet Chelonians in Europe According to Host Species

Rachel E. Marschang, PD Dr. med. vet., DipECZM (Herpetology), FTÄ Mikrobiologie,
Christoph Leineweber, Dr. med. vet.,
Elisabeth Müller, Dr. med. vet., DipECVM

Session #2627

Affiliation: From Laboklin GmbH & Co. KG, Bad Kissingen, Germany.

Note: Abstract format selected by author(s). See Front Matter for description of conference proceedings manuscripts submitted as scientific abstracts.

Abstract: Herpesviruses are well known pathogens in some species of chelonians, but there is still much work to be done to understand the role of these viruses in various host species. Herpesvirus detection rates in samples from chelonians submitted to a commercial laboratory between 2016 and 2020 were analysed according to host species. A total of 4,797 samples were analysed using a pan-herpesvirus PCR. These included samples from 3,475 Testudinidae, 176 Emydidae, 49 Geoemydidae, 10 Cheloniidae, 8 Pleurodira, 8 Kinosternidae, 4 Platysternidae, 3 Chelydridae, and 2 Carettochelydae, as well as 1,062 samples that could not be assigned to a specific host family or suborder. Herpesviruses were detected in a total of 312 (6.5%) chelonians. Detection rates differed significantly depending on host species ($P < 0.0001$). The highest detection rates were found in Horsfield's tortoises (*Testudo horsfieldii*) (54/361, 14.96%), radiated tortoises (*Astrochelys radiata*) (17/121, 14.05%), Chaco tortoises (*Chelonoidis chilensis*) (10/35, 28.57%), and box turtles (*Terrapene* spp.) (7/30, 23.33%). The results of this retrospective study indicate that there are differences in the epidemiology of herpesviruses depending on the host species, although many other factors including virus type, life history, and husbandry likely also play a role.

Section 28

Squamata

Sue Horton, DVM;
Shane Simpson, BVSc(Hons), GCM(VP)
Session Chairs

Intestinal Torsion in a Bearded Dragon (*Pogona vitticeps*)

Jessica May Hornby, BVetMed, PGDipVCP, MRCVS,
Joanna Hedley, BVM&S, DZooMed (Reptilian), DipECZM (Herpetology), MRCVS,
Simon Spiro, MA, VetMB, MVetMed, DPhil, DACVP, FRCPath, MRCVS

Session #2570

Affiliation: From Beaumont Sainsbury Animal Hospital, Royal Veterinary College, London, England (Hornby, Hedley), Zoological Society of London, Regent's Park, London, England (Spiro).

Note: Abstract format selected by author(s). See Front Matter for description of conference proceedings manuscripts submitted as scientific abstracts.

Abstract: A female spayed bearded dragon (*Pogona vitticeps*) of unknown age presented with a 48-hour history of anorexia and lethargy. On presentation the patient had a black beard with sunken eyes and full coelomic cavity on palpation although no specific thickening or masses were noted. Diagnostic investigations were conducted in the form of radiographs and ultrasonography with blood sample collection and packed cell volume assessed. These revealed a distended thin-walled fluid filled tubular structure within the ventral mid coelom and a packed cell volume of 3%. The owner elected euthanasia based on the patient's worsening demeanor and severe anemia well below the reference range.[1] A post-mortem examination with histology revealed a torsion of the gastro-intestinal tract through the mesentery with marked color change to the intestines. This was suspected to be the likely cause of death. To the authors' knowledge this is the first reported case of this presentation in this species.

References

1. Howard JG, Jaensch S. Haematology and plasma biochemistry reference intervals in wild bearded dragons (*Pogona vitticeps*). *Austr Vet J.* 2021;99(6):236-241. doi.org/10.1111/AVJ.13060.

Bacterial and Fungal Microbiome of the Bearded Dragon

Krista Keller, DVM, Dipl ACZM, and
Janina Krumbeck, PhD

Session #2621

Affiliation: From Wildlife Epidemiology Laboratory, College of Veterinary Medicine, University of Illinois at Champaign-Urbana, Urbana, IL 61801 USA.

Note: Abstract format selected by author(s). See Front Matter for description of conference proceedings manuscripts submitted as scientific abstracts.

Abstract: Recent literature has supported the tremendous role that microbiome plays in both healthy and diseased states in several species. Relatively little work has focused on the microbiome of reptiles at this time. Ten healthy adult bearded dragons had samples taken from their oral cavity, skin and cloaca. After DNA extraction, next-generation DNA sequencing was performed and the sequence compared to sequence databases. In total, 647 different bacterial species in 306 different genera were detected. The oral microbiome was dominated by *Brassica napus* with 12.3 % relative abundance, *Devriesea agamarum* (6.6%), and *Staphylococcus saprophyticus* (4.3%). The cutaneous microbiome was dominated by the species *Kocuria indica* (9.4%), a species from the order Clostridiales (8.1%), and a species from the genus *Cellulosilyticum* (4.7%). The cloacal microbiome was dominated by two species from the genus *Corynebacterium*, *Corynebacterium* sp. (15.5%) and *Corynebacterium xerosis* (10.8%), and *Brevibacterium otitidis* (9.2%). The mycobiome showed 239 different species from 160 genera. The most abundant species in the oral samples were not able to be identified at this time (42.1%), a species from the genus *Kazachstania* (7.0%), and *Hyphopichia burtonii* (6.1%). The cutaneous mycobiome was dominated by a species from the fungal kingdom (20.1%), *Hyphopichia burtonii* (18.7%), and *Candida sake* (13.9%). The cloacal mycobiome was characterized by the highest relative abundance of *Candida tropicalis* (13.3%), *Kazachstania* sp. (11.6%), and *Candida sake* (11.1%). Future work should include larger numbers of bearded dragons and compare diseased to healthy individuals to determine the role that microbiome may play in diseased states.

Prevalence and Risk Factors Associated with Hepatic Lipid Changes in Bearded Dragons (*Pogona vitticeps*)

Trinita Barboza, DVM, DVSc, Leonardo Susta, DVM, PhD, DACVP, Drury Reavill, DVM, DACVP, DABVP (Avian Practice)^, Hugues Beaufrère, DVM, PhD, Dipl ACZM, DABVP (Avian Practice), DipECZM (Avian)

Session #2544

Affiliation: From Department of Clinical Sciences, Cummings School of Veterinary Medicine, Tufts University, North Grafton, MA 01536, USA (Barboza), Department of Pathobiology, Ontario Veterinary College, University of Guelph, Guelph, Ontario, Canada (Susta), Department of Veterinary Medicine and Epidemiology, UC Davis School of Veterinary Medicine, Davis, CA 95616, USA (Beaufrère), ^Deceased (Reavill).

Note: Abstract format selected by author(s). See Front Matter for description of conference proceedings manuscripts submitted as scientific abstracts.

Abstract: Hepatic lipidosis is a common disease of captive, bearded dragons (*Pogona vitticeps*). Diagnosis, prevention, and treatment of this condition are challenging, as there is minimal information in the literature. Our study determined the prevalence and epidemiologic risk factors associated with hepatic lipid change grade and severity in bearded dragons submitted for necropsy in two North American institutions. A total of 571 post-mortem cases were retrieved, and from each pathology report the demographic data (age, sex) and the list of final diagnoses were extracted. For each case diagnosed with hepatic lipidosis, the archived sections of liver were reviewed and the severity of lipid change was stratified using a standardized histologic grading system. Descriptive statistics were used to estimate the prevalence of each grade and severity class. Associations between grade and severity, as well as demographic data and concurrent diseases, were explored using ordinal logistic regression analysis.

On multiple logistic models, the occurrence of infectious disease and neoplasia were associated with decreased hepatic lipid change grade and severity, while the female sex and adult age were associated with a higher grade and severity. All other variables were not significantly associated with hepatic lipid changes. These results suggest that reproductively active females and adult bearded dragons are predisposed to increasing hepatic lipid changes, while those with an underlying disease process have reduced hepatic lipid accumulation and changes, possibly due to increased fat catabolism. Data in this study can serve to benchmark the prevalence of hepatic lipidosis in bearded dragons and allow further investigations.

Real-Life Ouroboros: Investigation and Treatment of Tail Biting Behaviour in a California Kingsnake (*Lampropeltis californiae*)

Becky Mott, BSc(hons), BVetMed, and
Joanna Hedley, BVM&S DZooMed (Reptilian) DipECZM (Herpetology)

Session #2661

Affiliation: From Vets4Pets Newmarket, Studland Retail Park, Fordham Road, Newmarket, CB8 7SX, UK (Mott) and RVC exotics service, The Royal Veterinary College, Royal College Street, London, NW1 0TU, United Kingdom (Hedley).

Note: Abstract format selected by author(s). See Front Matter for description of conference proceedings manuscripts submitted as scientific abstracts.

Abstract: Currently, anecdotal reports and discussion of the ouroboros, the ancient symbol of the snake devouring its own tail, vastly outnumber scientific literature regarding tail biting behaviour in snakes. This presentation discusses the first case report of investigations and treatment of episodic tail biting in an 11-year-old male California kingsnake (*Lampropeltis californiae*) spanning 3 years, and discusses novel gabapentin and deslorelin implant use in this species. Initial investigation of tail biting behaviour in April 2019 revealed displaced coccygeal vertebrae, and a tail tip amputation resolved the behaviour until May 2020. Computed tomography then revealed coccygeal vertebral subluxation and suspected nerve entrapment, and a temporary response to an ultrasound-guided nerve block suggested pain as the possible inciting cause. Gabapentin was trialled at 10 mg/kg PO once, which did not prevent tail biting, although a long-lasting sedative effect was observed. A second tail amputation resolved the behaviour for 11 months until May 2021. The seasonality of the behaviour suggested a possible hormonal cause, and a deslorelin implant (Suprelorin 4.7 mg) was placed in June 2021 after obtaining baseline values for testosterone and cortisol. Multi-modal analgesia was resumed, including gabapentin 10 mg/kg PO q72h and tramadol 10 mg/kg PO q48h, alongside taping the mouth closed for 6 weeks post-implantation to prevent self-mutilation as the tail biting had intensified. Repeat testosterone and cortisol values initially decreased from baseline before increasing in December 2021 and again in January 2022, prompting repeat deslorelin implant (Suprelorin 9.4 mg) placement in January 2022. No further tail biting behaviour has been noted.

Section 29

Amphibians

Paul Gibbons, DVM, DABVP (Reptile and Amphibian Practice);
J. Jill Heatley, MS, DVM, DABVP (Avian Practice), DABVP (Reptile and Amphibian Practice), Dipl ACZM
Session Chairs

Cutaneous Iridophoroma in an Argentine Horned Frog (*Ceratophrys ornate*)

Charlotte Cournoyer, DVM,
Patrick Sullivan, DVM, DABVP (Avian Practice),
Pamela J. Mouser, DVM, MS, DACVP

Session #2667

Affiliation: From Avian Exotic Service, MSPCA Angell West, 293 2nd Ave, Waltham, MA 02451, USA (Cournoyer, Sullivan), Pathology Service, Angell Animal Medical Center, 350 S Huntington Ave, Boston, MA 02130, USA (Mouser).

Note: Abstract format selected by author(s). See Front Matter for description of conference proceedings manuscripts submitted as scientific abstracts.

Abstract: A 14-year-old Argentine horned frog (*Ceratophrys ornate*) presented for evaluation of a chronic, cutaneous mass. A white, semi-firm raised mass was present on the right cranial dorsum. Excisional biopsy was performed, followed by a second surgical removal 2 months later due to local recurrence. Fine needle aspiration cytology and histopathology were consistent with an iridophoroma. No recurrence has been noted greater than 1 year at the time of submission. While chromatophoromas are relatively common cutaneous neoplasms in reptiles and amphibians, iridophoromas are rarely reported in amphibians.[1-24] To the author's knowledge, this is the first report of an iridophoroma in an Argentine horned frog.

References

1. Atkins VM, Madden V, Churchill S, Godfrey V. Cutaneous iridophoroma in a spadefoot toad (*Spea* spp.). *Proc Am Coll Vet Pathol Am Soc Vet Clin Pathol.* 2013.

2. Bagnara JT, Hadley ME. The control of bright colored pigment cells of fishes and amphibians. *Am Zool.* 1969;9:465-478.

3. Bagnara JT, Taylor JD, Hadley ME. The dermal chromatophore unit. *J Cell Biol.* 1968;38:67-79.

4. deBrot S, Sydler T, Nufer L, Ruetten M. Histologic, immunohistochemical, and electron microscopic characterization of a malignant iridophoroma in a Dwarf Bearded dragon *(Pogona henrylawsoni)*. *J Zoo Wildl Med.* 2015;46:583-587.

5. Elkan E. Three different types of tumors in Salientia. *Cancer Res.* 1963;23:1641-1645.

6. Heckers KO, Aupperle H, Schmidt V, Pees M. Melanophoromas and iridophoromas in reptiles. *J Comp Pathol.* 2012;146:258-268.

7. Hopewell E, Harrison SH, Posey R, et al. Analysis of published amphibian neoplasia case reports. *J Herpetol Med Surg.* 2020;30(3):148-155.

8. Jacobson ER, Ferris W, Bagnara JT, Iverson WO. Chromatophoromas in a pine snake. *Pigment Cell Res.* 1989;2:26-33.

9. Khudoley VV, Eliseiv VV. Multiple melanomas in the axolotl (*Ambystoma mexicanum*). *J Nat Cancer Inst.* 1979;63(1):101-103.

10. Leach MW, Nichols DK, Hartsell W, Torgerson RW. Radiation therapy of a malignant chromatophoroma in a Yellow rat snake (*Elaphe obsoleta quadrivittata*). *J Zoo Wildl Med.* 1991;22:241-244.

11. Lewis N, Martinson S, Wadowska D, Desmarchelier M. Malignant mixed chromatophoroma with cutaneous, pulmonary and testicular metastases in a veiled chameleon (*Chamaeleo calyptratus*). *J Herpetol Med Surg.* 2015;25:16-20.

12. Ligon RA, McCartney KL. Biochemical regulation of pigment motility in vertebrate chromatophores: a review of physiological color change mechanisms. *Curr Zool.* 2015;62(3):237-252.

13. Menger B, Jahn S, Allmeling C, et al. Therapy of a melanosarcoma in an axolotl (*Ambystoma mexicanum*) by resection and skin grafting. [Thrapie eines malignen Melanosarkoms durch Resektion und Vollhaut-Transfer bei einem Axolotl (*Ambystoma mexicanum*)]. *Kleintierpraxis.* 2009;54(3):149-154.

14. Monahan CF, Meye A, Garner MM, Kiupel Matti. Gross, histologic, and immunohistochemical characteristics of cutaneous chromatophoromas in captive bearded dragons. *J Vet Diagn Invest.* 2021;33(5):932-938.

15. Munoz-Gutierrez JF, Garner MM, Kiupel M. Cutaneous chromatophoromas in captive snakes. *Vet Pathol.* 2016;53:1213-1219.

16. Pessier AP. Amphibia. In: Terio KA, McAloose D, St. Leger J, eds. *Pathology of Wildlife and Zoo Animals.* Cambridge, MA: Academic Press; 2018:921-952.

17. Rousselet E, De Mello Souza CH, Wellehan JFX, et al. Iridophoroma in a green iguana (*Iguana iguana*). *Vet Clin Pathol.* 2017;46(4):625-628.

18. Schmidt-Posthaus H, Pugovkin D, Wahli T. Iridophoroma in a European grayling (*Thymallus thymallus*) in Switzerland. *Bull Eur Assoc Fish Pathol.* 2005;25:36-39.

19. Singaravel V, Gopalakrishnan A, Vijayakumar R, et al. Occurrence of chromatophorma or chromatophoromatosis in *Sardinella longiceps* (Valenciennes, 1847) from Tamilnadu, southeast coast of India. *J Appl Ichthyol.* 2016;32:712-717.

20. Stolk A. Tumours of amphibians. VII. Erythrophoroma in *Dendrobates typographicus*. *Proc Akad Wet Amsterdam.* 1959;62C:381-389.

21. Stolk A. Tumours of amphibians. VIII. Guanophoroma in the frog *Hyla arborea meridionalis*. *Proc Akad Wet Amsterdam.* 1959;62C:390-395.

22. Szydlowski P, Pawel Madej J, Duda M, et al. Iridophoroma associated with the lemon frost colour morph of the leopard gecko (*Eublepharis macularius*). *Sci Rep.* 2020;10:5734.

23. Tong LJ, Ong W, Hulst F, et al. Clinical, diagnostic, and pathological features of 2 cases of metastatic iridophoroma in a veiled chameleon (*Chamaeleo calyptratus*) and a red-barred dragon (*Ctenophorus vadnappa*). *J Exot Pet Med.* 2018;22:53-60.

24. Varga JF, Bui-Marinos MP, Katzenback BA. Frog skin innate immune defenses: sensing and surviving pathogens. *Front Immunol.* 2019;9:3128.

Use of Implants for Terbinafine Administration to Prevent Chytridiomycosis in Greater Sirens (*Siren lacertina*)

Anastasia E. Towe, DVM,
Rebecca Hardman, DVM, PhD,
Wesley Sheley, DVM, DACVP,
Joseph DeMarchi, MS,
E. Davis Carter, MS,
Debra L. Miller, DVM, PhD

Session #2691

Affiliation: From Department of Comparative and Experimental Medicine (Towe, Sheley) and the Department of Biomedical and Diagnostic Sciences (Miller), University of Tennessee, College of Veterinary Medicine, 2407 River Drive, Knoxville, TN 37996, USA, Department of Forestry, Wildlife, and Fisheries, University of Tennessee, Institute of Agriculture, 2621 Morgan Cir Dr, Knoxville, TN 37996, USA (Miller, Hardman, Carter, DeMarchi), Florida Fish and Wildlife Conservation Commission, 620 S. Meridian St, Tallahassee, FL 32399, USA (Hardman).

Note: Abstract format selected by author(s). See Front Matter for description of conference proceedings manuscripts submitted as scientific abstracts.

Abstract: Batrachochytrium dendrobatidis (Bd) is a chytrid fungus that infects amphibians and is present in habitats worldwide.[1] Bd chytridiomycosis has been documented in wild lesser sirens (*Siren intermedia*)[2] and some greater sirens (*Siren lacertina*) were positive for Bd upon arrival at the Gray-Miller Amphibian Disease Lab at the University of Tennessee, Knoxville. This study evaluated the efficacy and safety of an anti-fungal drug (terbinafine) along with a novel delivery method, a terbinafine-impregnated intracoelomic implant (Melatek LLC), in prevention and clearance of Bd infection in greater sirens. Four sirens received terbinafine implants and four received blank implants. The animals were anesthetized with benzocaine water baths, and implants were placed intracoelomically through a small incision. The sirens were exposed to Bd zoospores at one- and two-months post-implant placement. Blood was collected from the ventral caudal tail vein monthly for plasma terbinafine levels, and skin swabs performed weekly for Bd qPCR at the Gray-Miller lab. Animals with terbinafine implants showed increased levels of terbinafine in plasma; however, treatment did not prevent infection and the clearance rate of Bd infection was not different from control animals. These findings indicate that intracoelomic drug implants are a safe method for antifungal drug delivery in amphibians; however, terbinafine efficacy in preventing chytridiomycosis remains unclear.

References

1. Scheele BC, Pasmans F, Skerratt LF, et al. Amphibian fungal panzootic causes catastrophic and ongoing loss of biodiversity. *Science.* 2019;363:1459-1463.

2. Chatfield MW, Moler P, Richards-Zawacki CL. The amphibian chytrid fungus, *Batrachochytrium dendrobatidis*, in fully aquatic salamanders from southeastern North America. *PLoS ONE.* 2012;7(9):e44821.

Risk of Bacteremia Associated with Probiotic Treatment of *Batrachochytrium salamandrivorans*

Anastasia E. Towe, DVM,
Sreekmari Rajeev, BVSc & AH, PhD, Dipl ACVM, DACVP,
Debra L. Miller, DVM, PhD,
Wesley Sheley, DVM, DACVP,
E. Davis Carter, MS,
Joseph DeMarchi, MS, Molly Bletz, PhD,
Douglas Woodhams, PhD,
Matthew J. Gray, PhD

Session #2698

Affiliation: From Department of Comparative and Experimental Medicine (Towe, Sheley) and the Department of Biomedical and Diagnostic Sciences (Rajeev, Miller), University of Tennessee, College of Veterinary Medicine, 2407 River Drive, Knoxville, TN 37996, USA, Department of Forestry, Wildlife, and Fisheries, University of Tennessee, Institute of Agriculture, 2621 Morgan Cir Dr, Knoxville, TN 37996, USA (Miller, Carter, DeMarchi, Gray), Department of Biology, University of Massachusetts, 100 Morrissey Blvd. Boston, MA 02125, USA (Bletz, Woodhams).

Note: Abstract format selected by author(s). See Front Matter for description of conference proceedings manuscripts submitted as scientific abstracts.

Abstract: Batrachochytrium salamandrivorans (Bsal) is a recently described fungal pathogen that has caused declines of wild salamanders and been detected in captive populations in Europe. One potential preventative and/or treatment option for Bsal chytridiomycosis is the application of probiotic bacteria. Previous studies have shown that certain bacteria present on amphibian skin have anti-fungal properties. This study aimed to determine the safety and efficacy of using a probiotic bacteria isolated from the skin of newts (*Pseudomonas fluorescens*) to prevent and treat chytridiomycosis. Alternatively, some bacteria can become pathogenic when given the opportunity. Thus, a secondary aim was to determine if bacteremia plays a role in the pathogenesis of chytridiomycosis. Eastern newts (*Notophthalmus viridescens*) were exposed either to Bsal, probiotic bacteria, pathogenic bacteria (*Aeromonas hydrophila*), or a combination of Bsal and one of the bacterial species. They were monitored throughout the course of disease development and samples were collected for PCR, blood cultures, and histopathology. Our results indicate that Bsal loads on the skin measured by PCR were lower in some groups that were exposed to probiotic bacteria, but mortality ranged from 67-100% and did not differ among treatment groups. We also found evidence of bacterial infiltration into the blood and organs in animals that were exposed to Bsal in probiotic-exposed and -unexposed newts. Thus, our results suggest that the *P. fluorescens* isolate was not an effective probiotic, and future studies should examine the potential for probiotics to contribute to bacteremia and Bsal pathogenesis.

Section 30

Nutrition and Reproduction

Paul Gibbons, DVM, DABVP (Reptile and Amphibian Practice);
J. Jill Heatley, MS, DVM, DABVP (Avian Practice), DABVP (Reptile and Amphibian Practice), Dipl ACZM
Session Chairs

Dubia Roaches: A Nutritionally Appropriate Food for Insectivores Made Better by Gut Loading

Elise D. Barras, DVM,
Kimberly L. Boykin, DVM, MS,
Maria G. Aguilar, DVM,
Zoe N. Lex, BS,
Amy K. Bitter, BS,
Mark A. Mitchell, DVM, MS, PhD, DipECZM (Herpetology)

Session #2612

Affiliation: From Division of Laboratory Animal Medicine (Barras), Department of Veterinary Clinical Sciences, School of Veterinary Medicine, Louisiana State University, Skip Bertman Road, Baton Rouge, LA 70803, USA (Barras, Boykin, Aguilar, Lex, Bitter, Mitchell).

Note: Abstract format selected by author(s). See Front Matter for description of conference proceedings manuscripts submitted as scientific abstracts.

Abstract: Dubia roaches (*Blaptica dubia*), a feeder insect originating from Central and South America, are often fed to captive reptiles; however, they have a natural inverse calcium to phosphorous (Ca:P) ratio. Reptiles fed non-gut loaded insects are more susceptible to developing nutritional secondary hyperparathyroidism. The study's purpose was to determine whether commercial gut loading diets could be used to produce a positive Ca:P ratio in these insects. There were two phases to this study. In phase one, roaches were fasted for 24 hours and randomly divided into a control group ($n = 6$, baseline) and three experimental groups (12, 24, and 48 hours; $n = 6$ each). Each cohort (n) represented approximately 75-80 roaches (15-18 g). Roaches in the experimental groups were gut loaded (50 g) with a high-calcium dubia roach diet. The Ca:P ratios of the roaches were found to significantly increase at all three time points compared to baseline ($P < 0.0001$; all means > 1.4). For phase two, roaches were fasted (24 hours) and randomly divided into a control group ($n = 6$, baseline) and two experimental groups ($n = 6$ each). Roaches in the experimental groups were either fed a high calcium diet or low calcium diet for 24 hours. Cohorts in the high calcium diet group had a significantly increased Ca:P ratio ($P < 0.001$; mean 1.5) compared to the inverse Ca:P ratios of the control and low calcium diet group. These results confirm that a high calcium diet can increase the Ca:P ratio in dubia roaches; however, not all gut-loading diets can alter the Ca:P ratio.

Can Superworms (*Zophobas morio*) be Super Sources of Calcium for Captive Reptiles and Amphibians?

Amy Bitter, BS,
Kimberly L. Boykin, DVM, MS,
Mark A. Mitchell, DVM, MS, PhD, DipECZM (Herpetology)

Session #2618

Affiliation: From Louisiana State University, School of Veterinary Medicine, Department of Veterinary Clinical Sciences, Skip Bertman Drive, Baton Rouge, LA 70803, USA.

Note: Abstract format selected by author(s). See Front Matter for description of conference proceedings manuscripts submitted as scientific abstracts.

Abstract: Superworms (*Zophobas morio*) are one of the most popular feeder insects in the commercial pet trade because they are readily available, easy to maintain, and remain in their larval state for an extended period of time. Unfortunately, these insect larvae are deficient in calcium and have a natural inverse calcium to phosphorus ratio (Ca: P). Reptiles and amphibians fed a calcium deficient diet are susceptible to the development of secondary nutritional hyperparathyroidism. To limit the occurrence of this disease, it is important to correct the inverse Ca: P ratio. Dusting superworms with a calcium carbonate powder prior to being offered, or gut-loading the superworms, may help to correct the calcium deficiency; however, evidence-based research to test these approaches is lacking. The goal of this study was to assess whether a commercial gut-loading diet for mealworms (*Tenebrio molitor*) could be used to gut-load superworms. In this study, super worms were gut-loaded with a high calcium mealworm diet for 0, 24, and 48 hours. All super worms were analyzed for dry matter, moisture, calcium, and phosphorous content at each time point. Superworms fed the high-calcium diet were significantly ($P < 0.001$) more likely to have a higher (positive) Ca:P ratio at 24 hours (Ca:P: 1.1 ± 0.14) and 48 hours (0.95 ± 0.19) then the controls (0.31 ± 0.05). There was no significant difference ($P = 0.10$) in the Ca:P ratio between the 24-hour and 48-hour groups. These results confirm that evidence-based research is needed to determine the value of different commercial gut-loading diets to ensure they provide appropriate nutrition.

Calcium Percentage and Gut-Loading Time Impact the Nutritional Quality of House Crickets (*Acheta domesticus*)

Maria G. Aguilar, DVM,
Elise D. Barras, DVM,
Kimberly L. Boykin, DVM, MS,
Zoe N. Lex, BS,
Amy K. Bitter, BS,
Mark A. Mitchell, DVM, MS, PhD, DipECZM (Herpetology)

Session #2712

Affiliation: From Department of Veterinary Clinical Sciences, School of Veterinary Medicine, Louisiana State University, Skip Bertman Road, Baton Rouge, LA 70803, USA.

Note: Abstract format selected by author(s). See Front Matter for description of conference proceedings manuscripts submitted as scientific abstracts.

Abstract: House crickets (*Acheta domesticus*) are one of the most common feeder insects offered to insectivorous species; however, these invertebrates lack the appropriate calcium concentrations most vertebrates require. As a result, diseases related to hypocalcemia (e.g., secondary nutritional hyperparathyroidism) are still common. Gut loading crickets with calcium-rich diets has been described as an excellent option for achieving a positive calcium-phosphorus ratio (Ca:P). Unfortunately, there remains uncertainty regarding the most appropriate length of time needed for gut loading and the minimum percentage(s) of calcium required for a positive Ca:P. The purpose of this study was to measure the Ca:P of house crickets over time using four different percentages of calcium in a gut loading diet. We hypothesized that Ca: P ratios would be highest after 6 hours but decline thereafter. Additionally, we expected that calcium diets > 4% would generate a positive Ca:P. Four-week-old house crickets fasted for 24 hours were randomly divided into 69 groups weighing approximately 18 g. The crickets were divided into five experimental groups based on gut loading time (0, 6, 12, 24, and 48 hours) and offered a diet comprised of 2%, 4%, 6%, or 8% calcium. Crickets were humanely euthanized at the end of the study. The Ca:P was positive at all times for the 4-8% diets. Crickets were also found to have the highest Ca:P ratios at 6 hours; the ratio declined thereafter. These findings suggest shorter gut-loading windows and diets with < 8% calcium can be used to prepare prey for insectivorous animals.

Evaluation of Two Semen Extenders Under Refrigerated Storage in Eastern Indigo Snakes (*Drymarchon couperi*)

Alexandra K. Mason, DVM, MS, CertAqV,
James Bogan, DVM, MVS (Con Med), DABVP (Canine and Feline Practice), DABVP (Reptile and Amphibian Practice), CertAqV,
Nick Clark,
Jake Swisher, AA, AS,
Mark A. Mitchell, DVM, MS, PhD, DipECZM (Herpetology)

Session #2557

Affiliation: From Central Florida Zoo & Botanical Gardens, 3755 W Seminole Blvd, Sanford, FL 32771, USA (Mason, Bogan), Central Florida Zoo's Orianne Center for Indigo Conservation, 30931 Brantley Branch Road, Eustis, FL 32726, USA (Clark, Swisher), Louisiana State University, School of Veterinary Medicine, Department of Veterinary Clinical Sciences, Skip Bertman Drive, Baton Rouge, LA 70803, USA (Mitchell).

Note: Abstract format selected by author(s). See Front Matter for description of conference proceedings manuscripts submitted as scientific abstracts.

Abstract: Reptiles are highly susceptible to anthropogenic activities as a result of narrow geographical ranges and habitat specialization. The Eastern indigo snake (*Drymarchon couperi*) is a large, nonvenomous, colubrid listed as threatened due to habitat loss. There exists a need to develop assisted reproductive technologies for this species to enhance breeding success and conservation efforts. Spermatozoa preservation is an important tool in conservation programs and can aid in future artificial insemination attempts with cooled or cryopreserved spermatozoa. The objectives of this study were to characterize semen parameters of the Eastern indigo snake, including color, volume, concentration, sperm motility and morphology, and to determine the spermatozoa motility over time, extended in two commercial semen extenders, Institut National de la Recherche Agronomique (INRA) 96 and sperm washing medium (SWM), stored at 4 °C (32 °F). Eleven, captive-bred, adult, male Eastern indigo snakes housed at the Orianne Center for Indigo Conservation were randomly selected for this study. Semen was manually collected from each snake during the breeding season. Spermatozoa were evaluated for motility at time zero, prior to extension, and every 24 hours until the motility was 0. Collection success was variable, with eight semen samples exposed to INRA and seven to SWM; four snakes had two separate samples exposed to both extenders. Fifty percent (4/8) of INRA 96 samples had 50% motilities at 48 hours, while only 14% (1/7) of SWM samples had a 50% motility at 48 hours. Further study is needed to assess INRA 96 as a semen extender in this species.

References

1. US Fish and Wildlife Service. Recovery plan for the Eastern indigo snake first revision. Athens, GA: Georgia Ecological Services Field Office; 2019.

2. Hyslop NL, Stevenson DJ, Macey JN, et al. Survival and population growth of a long-lived threatened snake species, *Drymarchon couperi* (Eastern indigo snake). *Popul Ecol.* 2012;54:145-156.

3. Fahrig BM, Mitchell MA, Eilts BE, Paccamonti DL. Characterization and cooled storage of semen from corn snakes (*Elaphe guttata*). *J Zoo Wildl Med.* 2007;38(1):7-12.

Ovarian Dynamics and Plasma Hormone Cyclicity in Two *Cyclura* Species

Sean M. Perry, DVM, PhD, and
Mark A. Mitchell, DVM, MS, DipECZM (Herpetology)

Session #2671

Affiliation: From Mississippi Aquarium, Gulfport, MS 39532, USA (Perry), Department of Veterinary Clinical Sciences, Louisiana State University, School of Veterinary Medicine, Skip Bertman Drive, Baton Rouge, LA 70803, USA (Perry, Mitchell).

Note: Abstract format selected by author(s). See Front Matter for description of conference proceedings manuscripts submitted as scientific abstracts.

Abstract: Iguanas are the most endangered family of reptiles, with 77% categorized as threatened or near threatened. Further, *Cyclura* is the most endangered reptile genus, with all 12 species considered threatened. Therefore, it is vital that we develop assisted reproductive technologies (ART) for *Cyclura* spp. to enhance their conservation efforts. Investigation into the fundamental physiology behind reproduction is necessary to develop ART. In Grand Cayman rock iguana hybrids (*Cyclura lewisi nubila*) and rhinoceros rock iguanas (*Cyclura cornuta*), male gonad and hormone cyclicity has been characterized previously; however, female reproduction has only been investigated at single time points for ecological studies. The goals of this study were to serially evaluate ovarian dynamics by evaluating follicle number, follicle structure, and follicular size in addition to measuring plasma estradiol, progesterone, and testosterone concentrations in captive Grand Cayman rock iguana hybrids and rhinoceros rock iguanas. A prospective longitudinal study was performed in eight female iguanas from each species during their reproductive seasons in southern Florida (January-August). The iguanas were manually restrained for blood collection from the ventral tail vein and ultrasounded using a (Butterfly VetIQ, Butterfly Inc.). Ovarian ultrasounds and blood were collected monthly. Plasma was used to measure estradiol, progesterone, and testosterone concentrations. A validated ELISA was used to measure hormones.

Section 31

ARAV Masterclasses

Kathleen Heinz-Taheny, DVM, DACVP;
Sean Perry, DVM, PhD;
Kimberly Wojick, DVM, Dipl ACZM
Session Chairs

Reptile Husbandry: Let's Cover the Basics

Grayson Doss, DVM, Dipl ACZM

Session #2527

Affiliation: From the Department of Surgical Sciences, University of Wisconsin-Madison, School of Veterinary Medicine, Madison, WI 53706, USA.

Abstract: High-quality husbandry is paramount for captive reptiles. Many commonly encountered disorders are related to poor husbandry. Unfortunately, there are few accurate husbandry resources readily available to owners and improper husbandry is a widespread reality with reptile pets. Because veterinarians routinely encounter pet reptiles, they must be prepared to provide accurate husbandry advice and intervene when adjustments are required. It is important to address husbandry issues as soon as possible to prevent disease, or at least as soon as possible, as therapy and supportive care may be negatively affected (or even ineffective) if husbandry remains suboptimal. Misconceptions or lack of knowledge regarding environmental temperature, humidity, diet and supplementation, water access, sanitation, lighting and cage substrate are common.

Introduction

Sufficiently replicating a natural habitat and diet for many reptiles can be challenging. Some species may seemingly thrive in basic enclosures while a similar approach would be detrimental for others. At a minimum, caretakers should be aware of the dietary preferences, preferred optimum temperature zone (POTZ), ultraviolet B radiation (UVB) requirement and humidity needs of the captive species in question. Additionally, proper sanitation and quarantine procedures should be implemented. These major husbandry categories are where mistakes are commonly made. Because of the vast array of reptile husbandry products available to consumers, having the client bring photographs or item inserts from purchased products is invaluable for determining if there are husbandry issues that need to be addressed. Also, a smartphone photograph or video of the enclosure can help veterinarians determine if the enclosure meets the pet's husbandry needs.

Minimizing Stress

Minimizing stress is a major goal of proper reptile husbandry. By decreasing environmental stress, physiologic stress, and subsequently the risk of disease, is lessened. For new pets, handling and feeding should be postponed for at least several days to allow the reptile to become acclimated to the new environment. Species that are high stress or easily injured (e.g., day geckos, chameleons) or potentially dangerous (e.g., crocodilians) should not be regularly handled; however, many reptile species will tolerate handling for short periods (e.g., bearded dragons, some small python and chelonian species).

Housing

Choosing appropriate housing among the myriad of options can be challenging. One must know the natural behavior and lifestyle of the pet species in question to select the appropriate enclosure size, orientation and materials as well as substrate and furniture. For example, enclosures designed for arboreal animals are rarely suitable for terrestrial or fossorial species and choosing a glass aquarium for a species requiring excellent ventilation (e.g., chameleons) is ill-advised. Oftentimes, cage substrate and furniture are intentionally selected based on the natural climate or habitat of the species (e.g., tropical climate for iguanas, arid climate for leopard geckos).

Enclosure furniture can range from utilitarian to naturalistic. Utilitarian style is typically easier to clean albeit less aesthetically pleasing. Elaborate, naturalistic enclosures are costly to build and are not easily sanitized but may provide more enrichment for the occupant.

Substrate

Substrate, or the ground covering of the enclosure, must be safe for the pet species in question, as well as easy to clean. Some substrates may also contribute to the humidity level of the enclosure or can be an ingestion (and subsequent impaction) risk. Unfortunately, clients often choose the substrate based on aesthetics rather than other, more important aspects like ease of sanitation or ingestion prevention. All substrates made of particles small enough to be eaten should be considered an ingestion risk. This includes gravel and river rock in semi-/aquatic species. Some fossorial species (e.g., sand and rosy boas) will need a substrate that safely permits burying. Paper and reptile 'carpet' substrates are easy to clean with less ingestion risk than sand or coconut by-products, but are less attractive. Pine and cedar wood shavings should be avoided due to irritating oils and dust. Aspen bedding can be used successfully with many snake and some lizard species but can be an ingestion risk.

Furniture

Hiding areas are essential furnishings in reptile enclosures. They can help decrease stress and can provide an area of increased humidity (e.g., humidity box). Hiding areas should be present in different thermal areas so the animal does not have to choose between seclusion and thermoregulating. Live plants are aesthetically pleasing but require great lighting to thrive and may be consumed or dug-up by inhabitants.

Lighting

Maintenance of a normal photoperiod is important for normal behavior. A 12hr light:12hr dark approach works for most species. Strong day spectrum lighting (5500-6500K ideal) necessary for most thriving, live plants. Most reptiles need light with wavelength in the ultraviolet spectrum. Ultraviolet-A (UVA; 320-400 nm wavelength) is important for behavioral regulation while ultraviolet-B (UVB; 280-320 nm wavelength) is important for photochemical synthesis of 1,25-hydroxyvitamin D_3 (calcitriol, the active form of Vitamin D). Reptiles obtain vitamin D_3 in two main ways: either primarily through their diet or through endogenous synthesis. Access to UVB radiation is required if the primary method of producing vitamin D_3 is endogenously. Herbivorous reptiles need exposure to UVB to produce endogenous vitamin D_3. Carnivores and omnivores should obtain enough vitamin D_3 through diet and exposure to UVB was historically not recommended; however, recent research has demonstrated that these species have higher circulating 25-hydroxyvitamin D (calcidiol) when exposed to UVB radiation and it may have physiologic and behavioral benefits in captive reptiles. Care should be taken not to submit species not naturally exposed to high levels of UVB to bulbs that produce a large amount of this type of radiation. For example, nocturnal gecko species should not be routinely subjected to UVB-emitting bulbs designed for desert-dwelling species that bask in the mid-day sun as this could lead to skin or ocular issues. Lighting goals for captive reptiles should aim to provide both UVA to simulate a natural photoperiod and low to high levels (depending on the species) of UVB for either endogenous vitamin D_3 synthesis and/or behavioral well-being.

There are many different types of light bulbs, and most do not produce UVB but may produce UVA radiation. Because of this, it is important to identify the type, age, and brand of the bulb when discussing husbandry with clients. Asking clients to bring information about the lighting they are using to the appointment can be invaluable.

Use of heat bulbs marketed for nighttime use, often labeled as night, nocturnal or moon bulbs are not recommended because reptiles can actually see the light spectrums produced, disrupting their photoperiod.

UVA radiation is produced by many commercially available heat lights marketed for pet reptiles. In particular, incandescent, metal halide, mercury vapor, and halogen bulbs typically produce UVA light. LED and bulbs meant for nocturnal use do not emit UVA radiation.

Major types of UVB-emitting bulbs include fluorescent, mercury vapor and metal halide. The latter two also produce significant heat suitable for reptile basking. UVB-emitting fluorescent bulbs do not have significant heat output and most types emit UVB over short distances. T5-type fluorescent bulbs can project UVB much farther than other types of fluorescent tubes. Currently, commercially available LED lights do not produce relevant levels of UVB radiation for use with captive reptiles. For reptile keepers, knowledge of Ferguson Zones[1] and use of a digital ultraviolet index meter (e.g., Digital UV Index Radiometer, Zoo Med Laboratories Inc., San Luis Obispo, CA, USA) is ideal. Important discussion points with owners regarding UVB-emitting bulbs include the following:

- Expiration date
- Distance from pet
- Material between bulb and pet
- Location in enclosure

Clients should abide by manufacturer bulb distance guidelines and expiration dates and ideally confirm these recommendations with an ultraviolet index meter. Even though a UVB-emitting bulb still produces visible light does not mean it is still generating UVB radiation. Any material between the reptile and the bulb can decrease UVB radiation reaching the animal. Glass and plastic completely block UVB. Even mesh screen significantly reduces the amount of UVB transmitted. Reflectors are recommended to increase transmission. If the UVB-emitting bulb is not also used for basking (e.g., fluorescent bulb) then it should be placed in area commonly utilized by the reptile (e.g., favorite basking spot) to ensure exposure. The best source of UVA and UVB radiation is exposure to natural sunlight.

Temperature

The enclosure temperature should be based on the POTZ for the species. Goals for providing heating include the following:

- Provide a proper thermal gradient during both day and night
- Provide sufficient ambient light during the day and sufficient darkness at night

A thermal gradient should encompass the range of temperatures that comprise the POTZ. For example, the POTZ for the inland bearded dragon (*Pogona vitticeps*) is 25-40 °C (77-104 °F). The extremes of this range would correspond with the coolest (e.g., hide area) and hottest (e.g., basking) areas of the enclosure and there should be no areas in the enclosure outside of this range of temperatures, regardless of time of day or night. When designing the heating layout, strive to create several microclimates in the enclosure so the reptile can easily thermoregulate.

Enclosure temperatures should be closely monitored through daily use of digital and infrared handheld thermometers. Various options exist for heat supplementation, including light producing heat bulbs and non-light producing under-tank heating pads and ceramic heat emitters. For light-emitting bulbs, coverage patterns include flood types, which transmit heat over a wide area and spot types that deliver heat in a more focused beam. A combination of heating methods is typically required to provide a proper thermal gradient during both day and night.

Humidity

An improper level of humidity increases the risk of infection and other disorders, like dysecdysis and constipation. While historically considered most important for tropical and subtropical species, the lack of provision of high humidity areas for arid reptile species has likely contributed to common problems like chronic dehydration and constipation.

Daily measurement of the enclosure humidity levels using a digital hygrometer are recommended.

Typically, an inadequate level of humidity is a more common problem than too much. There are two main ways to increase humidity in the enclosure:

- Supplementation
- Retention

Misting the enclosure walls and substrate with distilled or filtered water or use of a fogger or humidifier can supplement humidity. Addition of a water feature or water dish with a large surface area can also increase humidity through evaporation. Provision of a humidity box (e.g., plastic storage container filled with soaked sphagnum moss) is an easy way to provide access to high humidity levels in enclosures where humidity regulation is otherwise challenging. Retention of humidity can be improved by using a more moisture-retaining substrate (e.g., coir) and/or by partially covering open or screened areas with plastic cling wrap or sheets of plexiglass/acrylic; however, while covering areas can increase humidity it also decreases ventilation. If the enclosure is too humid, try decreasing or stopping supplementation or increasing ventilation.

Quarantine

Proper quarantine of new pets is essential, as it is the best or only 'treatment' for many reptile diseases (e.g., cryptosporidiosis, paramyxovirus, inclusion body disease). Veterinarians should be comfortable providing advice to clients regarding separating new reptile pets from the existing collection and monitoring quarantined animals for clinical manifestation of disease.

Nutrition and Supplementation

Incorrect information and misconceptions are rampant in regard to reptile nutrition. In particular, selecting produce for herbivores and omnivores and supplementation of calcium and multivitamins are common areas of confusion and controversy. Veterinarians (and clients) should understand that captive diets fall far short of replicating the diversity and quality of foodstuffs available to wild reptiles. Because of this, providing as much dietary diversity as possible is recommended, to decrease the risk of nutritional deficiencies. There are a handful of excellent resources on reptile nutrition available to veterinarians.[2]

Depending on the feeding strategy of the species (e.g., carnivore, omnivore, herbivore), captive diets usually consist of commercial formulated diets, live or pre-killed prey and plant material, or combinations thereof. While pelleted, commercial diets may be the easiest way to prevent deficiencies, palatability can be an issue.

Live rodents and small mammals should not be fed to reptiles due to ethical concerns and the potential for serious injury to the reptile. Crickets may also injure small reptiles if hungry. Frozen rodents, small mammals and birds are recommended over live prey and should be thawed slowly and completely in refrigeration and warmed before being fed. While adult rodents are a good source of calcium, pinkie rodents and day-old quail are naturally deficient in calcium and vitamin D due to non-calcified skeletons. Insects are also calcium deficient, necessitating calcium and multivitamin supplementation prior to feeding.

When feeding insects as any portion of the diet, they should be properly gut-loaded and dusted with calcium and multivitamin supplements before being offered to reptiles. Gut-loading diets should be high in calcium and contain active vitamin A and be offered to insects for at least 24 hours prior to feeding. Insects should be dusted immediately before offering to reptiles. Calcium supplements should be used two to three times weekly and a multivitamin (containing active vitamin A and carotenoids) every other week.

References

1. Baines FM, Chattell J, Dale J, et al. How much UVB does my reptile need? The UV-Tool, a guide to the selection of UV lighting for reptiles and amphibians in captivity. *J Zoo Aquar Res.* 2016;4:42-63.

2. Boyer TH, Scott PW. Nutrition. In: Divers SJ, Stahl SJ, eds. *Mader's Reptile and Amphibian Medicine and Surgery.* 3rd Ed. St. Louis, MO: Elsevier; 2019:201-223.

An Introduction to Axolotl Medicine and Surgery

Patrick Sullivan, DVM, DABVP (Avian Practice)

Session #2640

Affiliation: From Angell Animal Medical Center, 350 South Huntington Ave, Boston, MA 02130, USA.

Introduction

The axolotl (*Ambystoma mexicanum*), also known as the Mexican axolotl, is a large amphibian, from the order Caudata. This animal, which is now endangered, is native to only two freshwater lakes near Mexico City, Lake Xochimilco and Lake Chalco. The axolotl is a neotonic amphibian, meaning it does not complete metamorphosis, and therefore retains its external gills for its entire life. The most unusual thing about this animal is their ability to regenerate. Not only soft tissue, but entire limbs will grow back following amputation.

Husbandry

Axolotls can be kept in fresh to slightly brackish water, provided it's free of chlorine and chloramines, which are often found in tap water. These additives can cause skin and gill damage over time. Commercially available dechlorinating agents for aquarium fish can be used for axolotls. Filtered, deionized, or distilled water can be used, especially if creating a slightly saline environment. The water temperature should be closely monitored, and kept between 62-65 °F, although brief elevations up to 68 °F are usually tolerated. An aquarium cooler may be needed to maintain the enclosure in this range. The pH should be checked regularly and kept between 7-7.6. Water quality should be monitored as well and can be checked using commercially available aquarium water testing kits. A 10-g aquarium or larger should be provided, especially when keeping multiple animals together. Mixed-sex groups can be housed together, provided all animals are of similar size. Although not common, aggression between conspecifics can be seen. Providing multiple areas to hide and increasing the overall enclosure size will often times resolve the problem. Filtration should be provided to help maintain proper water quality, and remove waste material. Low power filtration is recommended, as high-power filtration often creates fast moving, turbulent water flow. This can cause stress to the animal, as well as skin irritation. The habitat should include areas to hide, which can help to reduce stress and promote normal behavior. Care should be taken to avoid offering anything that may be mistaken for a food item and swallowed. This means only providing items larger than the animal's head, thus ensuring it's not ingested. Lighting can be provided, but should be indirect, since these animals live in low light naturally.

The diet of the axolotl in captivity should consist of numerous types of insects, shrimp, blood worms, earth worms, and commercial fish pellets. Owners should attempt to offer a varied diet, but be cautious of obesity, which is common in captive animals. Juvenile animals can be fed daily, but adults can be reduced to 3-4 times a week. Offering live prey can increase the risk of exposure to parasites, and care should be taken to minimize this practice.

Anatomy

Respiration is primarily achieved through external gills. The gills appear as three projections, or rami, on either side of the head, and the caudal surfaces of each are covered with feather-like filaments. These filaments are where gas exchange occurs, and the animal can flex and extend each ramus to increase water flow across them. This action is increased when in water above the preferred temperature range, possibly due to a decrease in dissolved oxygen at warmer temperatures. Like other amphibians, cutaneous respiration also occurs. Axolotls have primitive lungs, and occasionally swim to the surface to gulp oxygen. A three chambered heart is present, and similar to other poikilotherms, the rate is heavily influenced by the external temperature. With few exceptions, axolotls are completely aquatic, and move through the water with the aid of a large, vertically oriented tail. The flat head, and tubular body, along with small peripheral limbs, enable rapid movements when needed, although most of the time, the small limbs are used to slowly ambulate on the bottom of the enclosure. The thoracic limbs have four digits, while the pelvic limbs have five digits. Axolotls are carnivorous, and have a short, simple GI tract. Small, blunt teeth are present in the oral cavity, but are mostly ineffective for tearing food items. Axolotls have poor eyesight, but a strong sense of smell, which aids in finding food. They are sexually dimorphic when mature (12-18 months of age). Males have a large swelling at the cloaca, referred to by hobbyists as the jelly gland.

Proper Restraint

Axolotls have very sensitive skin and should only be handled when necessary. Powder-free nitrile gloves should be used to securely restrain the patient for examination. Care should be taken to avoid holding onto the gills, or putting direct pressure on them, as this can damage the filaments. A fine mesh net can be used for restraint, as long as the limbs are not able to get caught in the netting.

Preventive Care

Routine exams are recommended, and should include a visual exam, followed by hands on palpation of the coelom and peripheral limbs. Palpation of the gills is likely to result in trauma, and should be performed on an as needed basis. An oral exam may be attempted, but care should be taken to avoid damaging the oral cavity while attempting to open it. Water testing can be performed at this time as well, especially if the owner is new to keeping this species.

Common Conditions

While axolotls are considered a hardy species in captivity, they are not immune to health conditions requiring veterinary care. While not a complete list, this section should help the clinician assess and treat a myriad of conditions that may be seen in practice. Signs of illness, while often non-specific, can involve hyporexia, edema, degeneration of the gills, lethargy, skin lesions or erythema, or ocular lesions.

Bacterial infections

The infections are often linked to improper husbandry, resulting in stress, and immune suppression. Organisms such as *Aeromonas* and *Pseudomonas* spp. are commonly identified as the causative agent for localized, or systemic infections, including septicemia. Skin scrapings of specific lesions, or collection of coelomic fluid and blood can be beneficial, and allow for cytology and bacterial culture. While not common, mycobacteriosis has been identified in this species. Overcrowding, mixing populations, and exposure to live feed may increase the risk for *Mycobacterium* infection. While symptoms may look the same as other bacterial infections, there is no effective treatment for mycobacteriosis, and it carries a zoonotic risk.

Bacterial infections can be treated in a number of different ways. After testing the water to ensure there are no elevations in waste material, often the next step is to lower water temperature to between 55-62 °F. Baths are often used as well, and can contain non-iodized salts, antibiotics, or antiseptics. Salt baths of up to 15 g/L can be given once every 12 hours for 10 minutes. One specific salt bath, known as Holtfreter's solution, can be used to both treat infection, and maintain electrolyte levels. Like other amphibians, topical application of antibiotics provides reliable absorption without the stress of prolonged handling or injecting. If indicated, injectable antibiotics can be given subcutaneously, or intracoelomically. Common antibiotics used for treatment include enrofloxacin (avoid repeated injections due to tissue trauma), amikacin, ceftazidime, metronidazole, and trimethoprim sulphamethoxazole.

Fungal infections

Several opportunistic fungal infections can affect axolotls, often causing a white, fluffy looking tufts on the skin. The most common of these is *Saprolegnia* spp, a water mold that is ubiquitous in the environment. This organism will colonize pre-existing skin lesions, creating cotton-like tufts which darken with time. The most concerning fungus for axolotls, as well as all amphibians, is *Batrachochytrium dendrobatidis*, the chytridiomycete fungus. Symptoms can include increased shedding or sloughing of the skin, but often death is the only symptom noted.

Much like bacterial infections, many fungal infections can be treated simply by correcting husbandry and reducing the water temperature slightly. Chytridiomycosis, among other fungal infections, has also been treated using an itraconazole bath combined with stringent cleaning of the environment.

Parasitic infections

Axolotls are susceptible to gastrointestinal parasites, as well as ectoparasites. Exposure to live prey items, such as fish, increases the risk of parasitism in captive animals. Common gastrointestinal parasites include *Hexamita* and *Opalina*. External parasites commonly seen include *Trichodina* and *Ichthyobodo*. The anchor worm (*Lernaea cyprinacea*) should also be considered, especially in enclosures where new animals were recently added. A wet mount of a fresh skin scraping or gill clipping is the best way to positively identify these parasites.

Salt baths, as well as metronidazole by mouth or in a bath, can alleviate a number of internal and external parasites. Recently a report by Stanley et al. was published where they used Lufenuron to treat *Lernaea cyprinacea*.

Gastric foreign body

Typically this is caused by ingestion of small stones or gravel in the enclosure, although pieces of fake plants, and other cage decor can create a risk. Small items may actually be regurgitated by the animal, but the majority will need to be removed with the use of forceps, or potentially an enterotomy. Care should be taken by owners to ensure that nothing in the enclosure is able to be swallowed.

Neoplasia

While not overly common, benign and malignant masses have been reported in the literature. A retrospective by Hopewell et al, described four cases of benign chromatophoroma, one neuroblastoma, and one malignant chromatophoroma. If indicated, cutaneous masses can be removed and left to heal by second intention. Application of topical antiseptics can help to reduce the risk of secondary infection during healing.

Trauma

While typically docile, overcrowding, housing different sized animals, and aggression during mating season can all lead to trauma secondary to bite wounds. Minor wounds can be cleaned with a topical antiseptic and left to heal, while more involved wounds may need sutures, or even amputation, if it involves a limb.

Anesthesia

Sedation and anesthesia may be indicated for a number of procedures in axolotls. Tricaine methanesulfonate (MS222) is the most commonly used anesthetic agent in amphibians, and is typically delivered via immersion. A wide dose range is published, but the author typically uses 1-2 g/L, allowing the patient to soak for up to 20 minutes, or until the righting response is absent. Due to the acidity of MS222, sodium bicarbonate should be used to buffer the solution to a pH of 7-7.1. Axolotls can be removed from the bath following loss of righting and placed on a wet surface for the procedure. Prolonged procedures may need application of the immersion solution to the patient during the procedure to maintain a surgical plane of anesthesia. Another option for sedation and anesthesia is alfaxalone. This drug has been used both as an immersion and an injectable drug in axolotls. As with MS222, irrigation of the gills may be needed for prolonged anesthesia. Drugs such as propofol and isoflurane have also been used to induce anesthesia, although both pose some degree of complication. Propofol does not provide reliable results, and isoflurane, while effective, is difficult to deliver safely. Topical application requires safe disposal of the drug following the procedure, and aeration of the drug into a bath produces a large amount of waste anesthesia gas, which is a risk to employees involved in the procedure.

Monitoring should always be performed when using sedation or anesthesia. Respirations should be counted, and the heart rate should be monitored using a doppler probe. The patient should be kept moist using appropriate water, or immersion solution.

Analgesia can be provided with injectable opioids such as butorphanol or buprenorphine, as well as NSAIDs given by mouth, or subcutaneously. Butorphanol can also be used in an immersion bath, for prolonged analgesia post-operatively.

Surgery

As we discussed, the axolotl has an amazing ability to regenerate tissue. This can be beneficial in regards to surgery, since amputated limbs will grow back to full size. The most common procedures performed on captive axolotls include gastric foreign body removal, limb amputation, and excisional biopsy of a mass. Gastric foreign body removal, depending on the object, may be performed using lubricated forceps in a sedated animal. If the object is too large to be removed via the esophagus, a enterotomy may need to be performed. This can be done under general anesthesia, with closure consisting of 2 layer closure with dissolvable suture material. Amputation is often required following trauma or severe infection involving a peripheral limb. Once anesthetized and prepped, the limb is amputated proximal to the affected area using Metzenbaum scissors or a scalpel blade. Topical antiseptics such as silver sulfadiazine or dilute povidone iodine can be used to keep the site clean, but sutures should be avoided, as they can hinder regeneration. Excisional biopsies can be performed similarly to other species, although closure is often not possible due to the tight skin of the axolotl. Much like amputations, the site can be treated topically while healing. Sutures can be placed if possible. In any surgical procedure, aftercare should consist of impeccable water quality, and isolation from conspecifics until fully healed.

References

1. Wright K, Whitaker B. *Amphibian Medicine and Captive Husbandry*. Malabar, FL: Krieger Publishing Co; 2001.

2. Carpenter JW. *Exotic Animal Formulary,* 5th Ed. St. Louis, MO: Elsevier; 2018.

3. LafeberVET. Axolotl. Available at: https://lafeber.com/vet/basic-information-sheet-axolotl/. Accessed May 27, 2022.

4. Loh R. Common disease conditions in axolotls. *Proc World Small Anim Vet Assoc World Congr.* 2015.

5. Gresens, Jill. An introduction to the Mexican axolotl (*Ambystoma mexicanum*). *Lab Anim.* 2004;33(9):41-46.

6. Del Valle JM, Eisthen HL. Treatment of chytridiomycosis in laboratory axolotls (*Ambystoma mexicanum*) and rough-skinned newts (*Taricha granulosa*). *Comp Med.* 2019;69(3):204-211.

7. Stanley J, MacHale J, Hedley J. Successful treatment of anchor worm (*Lernaea cyprinacea*) using lufenuron in the Mexican axolotl (*Ambystoma mexicanum*). *J Herpetol Med Surg.* 2021;31(2):107-110.

8. Hopewell Elizabeth, Harrison SH, Posey R, et al. Analysis of published amphibian neoplasia case reports. *J Herpetol Med Surg.* 2020;30(3):148-155.

9. Talbot B. 2018. The axolotl consult. *Proc Austr Vet Assoc Unusual Pet Avian Vet.* 2018.

Serving Up a Healthy Plate of Evidence-Based Nutrition Ensures Good Digestion for Reptiles

Kimberly L. Boykin, DVM, MS, and
Mark A. Mitchell, DVM, MS, PhD, DipECZM (Herpetology)

Session #2694

Affiliation: From Louisiana State University School of Veterinary Medicine, Skip Bertman Drive, Baton Rouge, LA 70803, USA.

Abstract: Reptile nutrition remains an important but understudied field. With more than 10,000 species of reptiles, there is much we need to learn if we hope to provide these animals with what they need to lead long lives. The purpose of this master class is to provide attendees a thorough review of our understanding of reptile nutrition, not through the lens of subjectivity and opinion, but using evidence-based research. Common examples of dogma are guide attendees through an interactive presentation that is meant to challenge them to apply science to the development of nutritional recommendations for their patients and clients.

Introduction

There are over 10,000 different species of reptiles, and as one might imagine, there are many different types of feeding strategies among these animals, including insectivory, carnivory, herbivory, and omnivory. Because of the diversity in the feeding strategies of these animals, and limited commercial resources, nutritional diseases remain a common problem in captive reptiles. An ideal diet for a captive reptile should mimic their natural diet and be as diverse as possible. Thus, even when limited options are available, the diversity provided can help offset some of the known potential deficiencies between foodstuffs.

Insectivores

Of the different feeding strategies, the group that tends to be the most challenging is the insectivores. True insectivores tend to be non-selective predators that, in the wild, balance their diet by eating a variety of insects. In the wild, these insects are exposed to different types of nutrition, including mineral rich soil, which allow them to transfer that nutrition to the reptile. The types of diets offered to captive insects are often focused on maintaining the insect and providing ample resources for them to reproduce. Because insects lack a skeleton, they have little natural need for calcium; thus, making them a poor source of this invaluable mineral for reptiles. Additionally, insects are naturally low in fat-soluble vitamins. Because of these natural deficiencies, it is important that we identify better methods of gut-loading insects prior to feeding them to captive reptiles. In captivity, there are some limitations regarding the availability of insects. The most common species available are the house cricket (*Acheta domesticus*), mealworm (*Tenebrio molitor*), superworm (*Zophobas morio*), black soldier fly larva (*Hermetia illucens*), dubia roach (*Blaptica dubia*), and the tomato hornworm (*Manduca sexta*). To date, most of the research focus has been on crickets and mealworms, with little known about the other species. Recent work in the authors' laboratory has been focused on identifying better methods of managing these prey insects in captivity to maximize their value to reptiles. A large portion of this presentation is to characterize best practices for maximizing insectivore nutrition using evidence-based research and to review the signs commonly associated with nutritional deficiencies in this group.

Carnivores

On the opposite end of the spectrum, carnivores tend to be one of the easier groups to feed. Whole prey items are relatively good sources of nutrients and can usually provide a completely balanced diet without the need for supplements (with the exception of thiamine and vitamin E for those species ingesting fish on a regular basis). Concerns associated with this feeding strategy are more focused on injuries from being offered live prey and obesity from being offered nutrient-dense meals too frequently without adequate exercise. Although many carnivores can subsist on a diet of only frozen mice, these reptiles should be offered a variety of prey items to better mimic the diet that would be available in their natural environment.

Herbivores

This group of reptiles can be quite challenging to feed correctly. Herbivorous species in their natural environment often consume a large variety of different plant species ranging from grasses and leaves to fruits and succulents. Plant availability in the wild can be ever-changing between seasons and even years, making copying their wild diet a complex proposition. Fruits and vegetables obtained from the grocery store have been selectively bred for human tastes and do not provide the same nutrients as plants found in the wild. Produce items should be chosen carefully and used in conjunction with other diet options such as commercial pellets, reptile-friendly plants, and grass hays to offer a better nutritional profile. Captive herbivores typically present for issues related to deficiencies in calcium, fiber, and some vitamins, improper protein intake (too low or too high), and excessive intake of simple carbohydrates. Common diseases include hepatic lipidosis, renal disease, nutritional secondary hyperparathyroidism, and hypovitaminosis D.

Omnivores

The omnivorous feeding style is utilized by a large number of reptile species. This feeding style is characterized by individuals that ingest both plants and animals with percentages of each varying across species and even life stage. Many juvenile reptiles will consume higher quantities of animal species in order to increase their protein and fat intakes while actively growing. Although this feeding strategy has the innate advantage of food variety, the issues that are seen in each of the other three feeding strategies are often seen in this group as well, particularly when the reptile develops certain food preferences. Calcium and vitamin deficiencies are often seen in animals ingesting large quantities of feeder insects. Hepatic lipidosis and obesity are often seen in reptiles selectively eating more animal protein than plant material.

Making Better Diets Through Evidence-Based Research

Crickets (*Acheta domesticus*)

Crickets are actually a good source of nutrition for reptiles. Their moisture, protein, and fat contents are appropriate and can provide essential building blocks for growth and reproduction. The greatest limitations associated with this species are that they have an inverse calcium: phosphorus ratio and low fat-soluble vitamin content.[1] The calcium: phosphorus ratio of crickets can vary based on the diet they are offered, but is rarely greater than 0.2:1 unless they are specifically gut-loaded with calcium. This severe inverse ratio can be problematic for young fast-growing reptiles or reproductively active females. Research has found that offering crickets a diet containing approximately 8% calcium can be used to generate a positive calcium: phosphorus ratio.[2] Unfortunately, because of the high mineral content, the crickets may stop consuming the diet, and thus maintain an inverse calcium: phosphorus ratio. Work done in our laboratory found that feeding the crickets high calcium diets for as little as 6 hours was sufficient to generate a positive calcium: phosphorus ratio, and that longer term feeding (24-48 hours) sometimes led to inverse ratios (unpublished data). This was likely due to the palatability of the diet. This is an important consideration for all gut-loading diets. For most pet owners, they offer the gut-loading diet and assume it works. The author can think of a number of cases where owners gut-loaded the crickets they were offering their pet, but that the reptile had clinical signs associated with secondary nutritional hyperparathyroidism. In some of these cases, the gut-loading diet was inappropriate (not high in calcium), while in others the diet was appropriate, but the crickets were not consuming sufficient quantities to generate a positive ratio. Regarding the former, it is important to recognize that not all commercially available gut-loading diets provide sufficient calcium (8%) to increase the calcium: phosphorus ratio. It is important to seek out the evidence regarding the diets that have been found to increase the ratio and include these in your recommendations to your clients. To minimize the likelihood of issues, the author also recommends feeding the crickets for 6-12 hours (and no more than 24 hours) prior to offering them to a reptile.

Mealworms (*Tenebrio molitor*)

Mealworms differ from crickets as a food package because they tend to be lower in moisture and much higher in fat content.[1] Reptiles fed a diet primarily consisting of mealworms (a common problem for some reptiles because of poor pet and owner compliance) tend to be obese and more likely to be chronically dehydrated. While this increased fat content may appear as a detriment, it is certainly of value under certain conditions. Much like in our own diets, fat is a good thing when provided in adequate concentrations. As more research studies have found, fat should not be restricted in a diet, but used in a balanced fashion. Thus, mealworms should be used as a component of an insectivore's diet to help diversify the nutrient profile of a diet. One of the issues that arise is that the reptile, much like a child, becomes selective in what it accepts. Some lizards prefer mealworms to other insects, leading to an unbalanced diet, especially related to calcium content. In the author's findings, mealworms tend to have the lowest calcium: phosphorus ratio (0.08:1).[1,3] In a reptile consuming mealworms as their primary diet, calcium deficiency is common. Fortunately, recent work with a commercial mealworm diet (Fluker Farms High-Calcium Mealworm Diet, Fluker Farms, Port Allen, LA) has found that the calcium: phosphorus ratio of mealworms (and superworms) can be converted to a positive ratio when offered the diet over 24 hours.[3] This is an important finding because it can help reduce disease in some animals that struggle with the diet they are offered.

Black solider fly larvae (*Hermetia illucens*)

Black soldier fly larvae have been identified as an important feeder insect because of their natural positive calcium: phosphorus ratio; however, this is based on their diet.[4] Initial reports focused on this positive ratio as an example of why these insects might serve as an important food source for captive reptiles. However, initial studies purely focused on measured mineral content of the black soldier fly larvae and not the digestibility of these insects. Much of the calcium found in the black soldier fly larvae is contained in the cuticle which oftentimes passes right through the gastrointestinal tract without being digested. When feeding these larvae, it is common for the feces to contain minimally digested larvae. In these cases, this positive calcium: phosphorus ratio may provide little benefit to the reptile.[5,6] To increase the value of these insects, puncturing the cuticle may assist with digestion. These larvae, like so many insects, are also low in vitamin A. Reptiles fed diets low in vitamin A can develop significant systemic illness. We have found that offering a vitamin A enriched diet to black soldier fly larvae can be used to increase their vitamin A content, and that this vitamin A can be transferred to the reptile being fed the larvae. This was proven by performing liver biopsies in leopard geckos (*Eublepharis macularius*).[6] It is important for us to continue to expand what we offer insects in their gut-loaded diet to ensure we are providing the best nutrition possible for captive reptiles.

Dubia roaches (*Blaptica dubia*)

Dubia roaches have continued to gain in popularity because of their longevity, ease of care, and ability to be contained in high back containers with little concern for escape. These roaches, like the crickets, are a balanced source of protein, fat, and moisture for reptiles; however, much like the other insects noted previously, dubia roaches also have an inverse calcium: phosphorus ratio.[7] Recent research in the authors' laboratory has found a commercially available high calcium diet (Fluker Farms High-Calcium Dubia Roach Diet) capable of increasing the calcium: phosphorus ratio of the roaches from a 0.11:1 to a > 2:1 ratio over 48 hours (unpublished data). These results will further refine our ability to develop a more balanced diet for captive reptiles.

Hornworms (*Manduca sexta*)

Tomato hornworms are a pest throughout much of their range. While looked about negatively by many, they, like many of the insects discussed here, are much appreciated by reptile caretakers. Unfortunately, there has been minimal information published regarding the nutritional value of these insects, and that which is published appears incorrect. Claims about high calcium concentrations are just not correct. Recent work in the author's laboratory has found that tomato worms have a significant inverse calcium: phosphorus ratio.[8] Initial attempts to correct this with a commercial diet (Repashy Superhorn, Oceanside, CA) were unsuccessful.[8] The greatest nutritional benefit that hornworms may provide captive reptiles is their moisture content. We found that the moisture content of these insects was approximately 85%.[8] Since reptiles derive much of their hydration through their diet, hornworms could be a good way to ensure the moisture levels are appropriate.

Commercial diets

The world of commercial diets for reptiles is ever-expanding with some diets being well-balanced and researched, while many others are not. As veterinarians, it is our job to direct owners to diets that have been specifically formulated to meet the needs of their reptile species. If well accepted by the reptile, commercial diets can be a useful tool to provide additional variety, nutrients of concern, and even moisture.

Research within the authors' lab has included analysis of semi-moist diets formulated for juvenile bearded dragons and omnivorous tortoises (Fluker's Crafted Cuisine diet line, Port Allen, LA). The juvenile dragon diet was shown to provide an appropriate Ca:P ratio and adequate levels of calcium to growing bearded dragons. This was demonstrated by measuring bone density with CT scans as compared to dragons receiving fasted crickets as their main source of food.[9] The semi-moist tortoise diet was shown to have higher calcium levels than the Mazuri dry tortoise pellets (Mazuri Exotic Animal Nutrition, St. Louis, MO), which resulted in mild elevations of serum calcium levels after only a few weeks. Both diets appeared to be well tolerated as long as they were served with leafy greens and did not lead to any clinical issues. Given the increased calcium levels, the authors suggest that the semi-moist diet could be a useful diet addition for young, growing tortoises or reproductively active females that may benefit from a higher calcium level in their diet.[10]

Another avenue of research into commercial diets involved trying to create more diet options for snakes. Using sausage casings, we were able to formulate an insect-based sausage diet for snakes. If accepted, the sausage diet could offer lower-calorie alternate protein sources that would reduce our dependency on frozen mice and reduce the potential for spreading *E. coli* and *Salmonella* strains. Our research found that the insect-based diet was capable of being digested and maintaining healthy, growing snakes, however, diet acceptance will likely be a challenge.[11]

Specific Nutritional Diseases

Nutritional disorders in reptiles commonly present as a chronic problem, and the diet is often times centered on limited food sources or human convenience. In most cases, deficient diets are comprised of limited numbers of food items and/or are not supplemented with calcium and vitamin powders.

Nutritional secondary hyperparathyroidism

Nutritional secondary hyperparathyroidism (NSHP) is the most common nutritional disorder reported in insectivorous reptiles. Often termed metabolic bone disease (MBD), it is defined as any metabolic defect that alters the morphology and functioning of bones. NSHP is usually related to low levels of calcium or excessive levels of phosphorus, which consequently bind to calcium and render it physiologically unavailable. Decreased calcium availability results in increased parathyroid activity and mobilization of stored calcium from the shell and bone cortices. Factors predisposing reptiles to the development of NSHP include: dietary deficiency of calcium and/or suitable vitamin D_3, inappropriate calcium: phosphorus ratio of the diet, lack of exposure to ultraviolet light (ultraviolet B radiation increases activation of vitamin D precursors and facilitates gastrointestinal absorption of calcium), dietary excess of protein during rapid growth periods, anorexia, or abnormal vitamin D_3 metabolism secondary to renal, hepatic, intestinal, or parathyroid disease. NSHP is commonly observed in rapidly growing juvenile reptiles and reproductively active females. Clinical signs consistent with NSHP vary depending on the age and species of the patient. The most common clinical finding in reptiles with NSHP include muscle tremors and fasciculations, seizures, pathologic fractures and acute death. A thorough history is required before a diagnosis of NSHP can be met. Diagnostically, radiographs and blood work can provide insight into the reptile's disease state. Radiography may reveal fibrous osteodystrophy and pathologic fractures. Low blood calcium levels are highly suggestive of NSHP, but calcium blood levels are frequently not low in cases of NSHP because of hyperparathyroid activity. It must be remembered that blood levels of calcium are not reflective of physiologically available levels of calcium. Ionized levels of calcium are more indicative of the availability of calcium, but, unfortunately, published reference levels are difficult to find in the literature. Treatment of NSHP is dependent upon the correction of inappropriate husbandry. An unsuitable calcium: phosphorus ratio of the diet should be corrected, the proper provision of ultraviolet light should be instituted and oral supplementation of calcium and vitamin D3 should be initiated if appropriate for the species. Supplemental calcium during the treatment period is also strongly recommended.

Hypovitaminosis A

Vitamin A is essential to many different systems in the body. Deficiencies can lead to squamous metaplasia and a loss of tight junctions between cells. This can lead to breakdowns in tissues that act as natural barriers, such as the skin and respiratory, gastrointestinal, and excretory tracts. Hypovitaminosis A is likely a common, underdiagnosed disease in insectivorous reptiles. As mentioned previously, insects are naturally low in vitamin A. As a result, and over time, susceptible animals can develop life-threatening issues. Diagnosis can be challenging, but a thorough history and examination are essential. Treatment should include supplementing vitamin A orally, and correcting deficiencies in the diet. Care should be taken with parenteral supplementation to reduce the risk of iatrogenic hypervitaminosis A.

Obesity

Some of the commercially available larval insects (e.g., mealworms) have a very high fat content. Because these prey items routinely comprise a large portion of a captive insectivore's diet (e.g., bearded dragon, leopard gecko, chameleon), these lizards are susceptible to an over-accumulation of fat. These species of lizards primarily store their fat in two large coelomic fat bodies. The fat bodies are anchored in the caudal coelomic cavity, and the cranial aspects of the fat bodies are unattached. In over-conditioned insectivores it is often possible to palpate the fat bodies. Obesity can lead to a number of health issues in captive reptiles, including dystocia and hepatic disease. To limit the likelihood of complications associated with obesity, it is important that clients provide an appropriate diet for their insectivores. The number of larval invertebrates should be limited in an animal's diet.

Summary

Captive insectivores can be a challenge for veterinarians because of limited options in captivity. While the prey insects we currently have available may be limited nutritionally, there are ways to improve the value of these prey insects. From gut loading them to increasing the overall diversity of prey being offered, veterinarians can make suggestions to help limit the potential for secondary nutritional diseases and ensure long lives for their reptile patients.

Disclosure statement: Fluker Farms provides research grants to fund veterinary and graduate student research. They have no input on studies performed.

References

1. Finke MD. Complete nutrient composition of commercially raised invertebrates used as food for insectivores. *Zoo Biol.* 2002;21(3):269-285.

2. Finke MD. Gut loading to enhance the nutrient content of insects as food for reptiles: a mathematical approach. *Zoo Biol.* 2003;22(2):147-162.

3. Boykin KL, Bitter A, Mitchell MA. Using a commercial gut loading diet to create a positive calcium to phosphorus ratio in mealworms (*Tenebrio molitor*). *J Herpetol Med Surg.* 2021;31(4):302-306.

4. Boykin KL, Mitchell MA. The value of black soldier fly larvae (*Hermetia illucens*) as a food source: A review. *J Herpetol Med Surg.* 2021;31(1):3-11.

5. Dierenfeld ES, King J. Digestibility and mineral availability of phoenix worms, *Hermetia illucens*, when fed to mountain chicken frogs, *Leptodactylus fallux*. *J Herpetol Med Surg.* 2008;18(3/4):100-105.

6. Boykin KL, Carter RT, Butler-Perez K, et al. Digestibility of black soldier fly larvae (*Hermetia illucens*) fed to leopard geckos (*Eublepharis macularius*). *PLoS ONE.* 2020;15(5):e0232496.

7. Kulma M, Plachy V, Kourimska L, et al. Nutritional value of three *Blattodea* species used as feed for animals. *J Anim Feed Sci.* 2016;25:354-360.

8. Mitchell MA, Windle D, Boykin KL. Determining the nutritional value of captive-bred and reared tobacco hornworms (*Manduca sexta*). *Proc ExoticsCon.* 2021;460.

9. Molitor L, Aschebrock S, Mitchell MA. You are what you eat! Evaluating a commercial bearded dragon diet. *Proc Annu Conf Assoc Reptilian Amphib Vet.* 2017;83.

10. Aschebrock S, Boykin KL, Ford C, Mitchell MA. What goes in, must come out! Evaluating the digestibility of commercial tortoise diets. *Proc ExoticsCon.* 2021;461.

11. Boykin KL, Butler-Perez K, Buck CQ, et al. Preliminary evaluation of a novel insect-based sausage diet for juvenile corn snakes (*Pantherophis guttatus*). *J Herpetol Med Surg.* 2020;30(3):129-136.

Comparison of Useful Analytes in Exotic Animal Practice: Proteins, Bile Acids and Lactate

J. Jill Heatley, DVM, MS, DABVP (Avian Practice), DABVP (Reptile and Amphibian Practice), Dipl ACZM

Session #2514

Affiliation: From Texas A&M University College of Veterinary Medicine and Biomedical Sciences, 660 Raymond Stotzer Pkwy, College Station, TX 77843, USA.

Introduction

Choice and interpretation biochemical analytes for exotic animal species is made challenging (and exciting) based on the wide variety of lifestyles, physiology and diversity of the many species often included within this category. For the purpose of this comparative review, I delve into some commonly used biochemical analytes, which differ based on reptile and amphibian species and assay. I also review some new information from the literature, which holds promise for diagnostic use in these species. While a clinical pathology finding should not create, but confirm, diagnosis for all species, the exotic animal diagnostic clinician and veterinarian often struggle with what is normal and if a reptile or amphibian believed healthy is in fact, suffering occult disease. The stoic nature of many of these undomesticated prey species which challenges us to correctly classify these patients to achieve best outcome for the animal and owner. Biochemical analytes which vary greatly in their diagnostic ability include the mammalian based enzymes, total protein, albumin and globulin (and multiple methods of assay), bile acids (and variety of assays).[1] In comparison, many chemical assays, such as the electrolytes, venous blood gases, lactate, glucose and multiple acute phase proteins hold promise as being more widely applicable and therefore able to discern health and disease in these species, particularly those that are non-mammalian.[2-5]

While many diagnostic analytes have assays that are suitable for reptiles and amphibians, assessment remains challenging based on an unknown range of the analyte in health or values tolerated by the species. Ionized calcium is an excellent example of advances in this area which demonstrate that the very basic building blocks, used by all species in the same basic form are excellent assays for diagnostic use. The challenge remains making published references ranges accessible and having the analyzer and pertinent conditions of the assay and animal reported for comparison (temperature for assay and ectotherm sample collection, UVB light exposure, age sex and diet or at specify what is any statistical patient or environmental conditions were tested and made no difference within the population tested. Therefore, for the remainder of this discussion I focus on the diagnostic possibilities and foibles of proteins, bile acids and lactate in reptiles and amphibians.

Proteins

Assays range from acceptable with analyzer-specific reference intervals, to completely incompatible with the reality of the patient (e.g., bromocresol green [BCG] in rabbits, turtles; however, refractometry remains a simple, cost-effective assay in those species without a stress glucose above 100, and a lack of lipemia and or vittelogenesis (e.g., male reptiles, non-reproductively active female reptiles, most rodents). Protein electrophoresis remains the gold standard but species-based differences, analyzer differences and the lack of absolute values (i.e., these are calculated from the BCG above) makes the basic proteins of albumin and globulins quite unreliable for these species. Comparison on the same analyzer from a previous value (trends) or from a health conspecific is recommended. Nonetheless, albumin is considered an excellent acute phase protein, which decreases in birds and reptiles during acute inflammation.

Bile Acids

These intermediate products of the heme breakdown pathway of the liver can indicate liver function; however, bile acids are highly variable among species. Nineteen bile acids occur from rat liver, plasma, bile, intestinal contents. Despite high variability of rat bile acids, this analyte is recommended to assess liver function in both the rat and mouse. Increased bile acid concentrations are presumed diagnostic of liver disease in both the rabbit and ferret as well as other exotic mammals, but few specific reports are available to provide data for this assumption. Thus, bile acids are considered a fairly reliable indicator of liver dysfunction most mammals and birds. Proof of bile acid utility beyond these species is minimal and remains to be investigated for many.

Clinical diagnostic utility of commercially available bile acid assays has been studied in tortoises, green iguanas, panther chameleon, leopard gecko and in aquatic turtles. Interpretation of this analyte is made challenging based on variation of the major bile acid with reptile and amphibians orders. Alligators, snake, frogs, and turtles have some particularly unusual bile acid structures.[1] Thus the value of commercial assays for bile acids in snakes, Crocodylia and many amphibian species remains unproven. Bile acid structures have been determined for about 200 reptiles and in about 80 amphibians based on investigations of evolutionary relationships.[1] For example in a recent study of four different tortoise species with commercial enzymatic assays for 3-alpha-hydroxy-bile acids, *Testudo* spp. *hermanii, graeca* and *Agronnemmys horsfieldii* values were similar (17.64 ± 27.26, 17.86 ± 15.37 and 20 ± 24.65) but distant from those of *Kinixis belliana* (6 ± 0.1).[2] While some values published suggest normal values of less than 60 μmol/L using commercial enzymatic assays for 3-alpha-hydroxy-bile acids, values of much less could indicate disease for some species. In the captive New Guinea snapping turtle, *Elseya novaeguineae*, bile acid top of 95% CI was higher for serum (0-44 mg/dl) compared to plasma (0-41 mg/dl).[3] For leopard geckos (*Eublepharis macularius, n =26*) bile acid reference intervals were 0.6-37.5 μmol/L.[4] For Panther chameleons (n = 27-40), age and sex affected bile acid reference intervals of 1.3-21.9, 1.7-23.2 for juvenile females and males and 1.0-10.6, 0.4-14.0 for adult females and males, respectively.[5]

Further investigations are needed to determine the effects of feeding upon bile acid values for most species; however, fasting status should be considered when evaluating bile acid concentrations. Bile acid diagnostic use in the green iguana appears indicated and fasting appears to affect this analyte in other reptiles;[6,7] however, while a significant post prandial increase in Bile acids occurred in the green iguana, a similar analyte decrease was not observed for the fasted RES. Pre-(48-hour fast) and post-prandial plasma 3α-hydroxy bile acids for 11 healthy male green iguanas, *Iguana iguana*. Pre-prandial bile acid concentrations (7.5 ± 7.8 μmol/L) increased 3 and 7.5 hours post-prandially to 33.3 ± 22.0 μmol/L and 32.5 ± 8.4 μmol/L, respectively. For 10 female red-eared sliders (RES), *Trachemys scripta elegans*, mean (SD) bile acid concentrations (μmol/L) were 7.35 (4.52)-10.04 (7.40). Fasting concentrations were lower but not statistically different 3.1 (2.3)-4.5 (5.4) at 24 hours and 2.2 (1.5) at 48 hours; however, in female RES, high concentrations of 27.6 and 66.6 μmol/L were associated with lipemia.[8]

Lactate

Lactate occurs in two isomers L-lactate, the form measured by point-of care (POC) devices, and created when the body undergoes anaerobic metabolism, and D-lactate which is generated by bacteria in the gut. L-lactate is produced at low concentrations by the red blood cells, brain, and skeletal muscles. In mammals, lactate is primarily metabolized by the liver, and secondarily by the kidneys; in amphibians, lactate is primarily metabolized by the muscles. Lactate concentrations may increase due to overproduction, failure of clearance, intense exercise, stress, sepsis, drug reactions, mitochondrial dysfunction and during critical emergencies or during anesthetic events, lactate can indicate hypoperfusion. A patient with a high or increasing lactate concentrations should be evaluated holistically, as a critically ill patient may have mixed origins of an elevated blood lactate concentration, such as could occur in a trauma patient with concurrent hepatic disease.

Point-of-care lactate meters are inexpensive, quick and easy to use, and require minimal blood sample sizes and training for use. The small sample size allows repeated sampling without patient harm. Comparison of benchtop analyzers and handheld lactate meters often demonstrate good agreement. In mammals, < 2.0 mmol/L is considered an acceptable blood lactate value, 2.0-4.0 mmol/L is hyperlactemia and > 4.0 mmol/L with a pH < 7.35 is lactic acidosis. These values do not appear to correlate to acceptable lactate values for reptiles or amphibians based on preliminary field studies a basic physiologic investigations.

As ectotherms, reptiles and amphibians likely rely on anaerobic metabolism more than endotherms, even during relatively short bursts of activity. Reptiles have diverse evolutionary adaptations to survive in anoxic conditions including slowed metabolic rates and ability to sequester and buffer lactate. Reptiles may have a much higher lactate tolerance and associated acidosis, than other taxa. Unfortunately, few studies in reptiles assess lactate concentrations. Disease, presumed hypoperfusion events such as cold stun, and activity, to include prolonged handling, can increase blood lactate concentrations of reptiles (Table 1).

Table 1. Lactate changes in reptiles and amphibians (Mean = x).

Species	Resting lactate (mmol/L)	Activity	Post lactate (mmol/L)	Reference
Iguana (*Iguana iguana*)	2.1	Intense treadmill run	16.5	17
Monitor (*Varanus exanthematicus*)	1.6	Intense treadmill run	20.5	17
Eastern copperheads (*Agkistrodon contortrix*)	10.3 x	Post recording morphometrics	12.7 x	18
Eastern rat snakes (*Pantherophis alleghaniensis*)	10.9 x	Post recording morphometrics	17.4 x	18
Galapagos lava lizards (*Microlophus bivittatus*)		Animals handled 5-10 minutes prior to sampling	17.08 x males 17.04 x females	19
Farmed Nile crocodiles (*Crocodylus niloticus*)	4.0 x	Electrostun versus noosed, noosed crocodiles struggle	10.2 x	20
Free-living Galapagos marine iguanas (*Amblyrhynchus cristatus*)	6.63 x	Captured and sampled within a few minutes		21
Galapagos green turtles (*Chelonia mydas*)	3.73 x	sampled within 5 minutes of capture		22

Painted turtles (*Chrysemys picta*) are unusual even among reptiles as they can tolerate lactic acid concentrations of up to 150-200 mmol/L. Freshwater turtles with hard shells may tolerate increased lactate concentrations more than soft shelled turtles (*Apalone spinifera*).[9] Painted turtles can remain submerged in anoxic conditions over-winter via multiple adaptations for lactate tolerance.[10] A slow metabolic rate reduces lactate production, high concentrations of bicarbonate in extracellular fluid, to include the pericardial space, and buffering and sequestering lactic acid in the shell and bones allow this remarkable behavior.[11] Disease and increasing capture time correlates with increased lactate concentrations in box turtles (*Terrapene carolina carolina*).[12] Lactate concentrations 2-3 days post stun event, increased significantly in cold stunned Kemp's Ridley sea turtles (*Lepidochelys kempii*), and were presumed linked to perfusion status.[13]

In *Bufo americanus*, lactate accumulates to 0.63 ± S.E. 0.06 mg lactate/g during a 10-minute bout of forced activity. The lactate fraction oxidized by aerobic metabolism (< 10% of total 14C lactate ~17%) is considerably less than the fraction of lactate oxidized by mammals of 20-90%.[14] Based on transport stress of the bullfrog (*Lithobates catesbeianus*), lactate (mmol/L , mean) increased in the animals at time 0 hours (29 compared to pre of 18), but recovered at 6 hours (15) and reduced to lower than pre time by 48 h after transport (9).[15] In amphibians lactate is mainly metabolized by the muscle, and not by the liver.[16]

Increased blood lactic acid concentrations generally result from activity and or hypoperfusion in reptiles and amphibians. While the inciting cause must be managed, supportive care is also appropriate. Symptomatic treatments include increasing oxygenation, inotropes, vasopressors, and resuscitative fluid therapy. In reptiles, administration of lactate free fluids may alleviate a high lactate burden of lactate. Treatment of lactic acidosis with sodium bicarbonate alone will not alleviate the primary problem, only the clinicopathologic symptom, and may be injurious as excessive bicarbonate administration can cause intracellular acidosis and exacerbate acidosis.

References

1. Hagey LR. *A Study of the Evolution and History of Bile Acids.* Available at: http://bile.ucsd.edu/. Accessed June 11, 2021.

2. Montesinos A, Martinez R, Jimenez A. Plasma bile acids concentration in tortoises: reference values and histopathologic findings of importance for interpretation. *Proc World Small Anim Vet Assoc.* 2002.

3. Anderson NL, Wack RF, Hatcher R. Hematology and clinical chemistry reference ranges for clinically normal, captive New Guinea snapping turtle (*Elseya novaeguineae*) and the effects of temperature, sex, and sample type. *J Zoo Wildl Med.* 1997;1:394-403.

4. Cojean O, Alberton S, Froment R, et al. Determination of leopard gecko (*Eublepharis macularius*) packed cell volume and plasma biochemistry reference intervals and reference values. *J Herpetol Med Surg.* 2020;30(3):156-164.

5. Laube A, Altherr B, Clauss M, Hatt JM. Reference intervals for bile acids and protein electrophoresis in plasma of captive panther chameleons (*Furcifer pardalis*): a first approach. *J Herpetol Med Surg.* 2018;28(3-4): 99-106.

6. McBride M, Hernandez-Divers SJ, Koch T, et al. Preliminary evaluation of pre-and post-prandial 3α-hydroxy bile acids in the green iguana, *Iguana iguana*. *J Herpetol Med Surg.* 2006;16(4):129-134.

7. Knotek Z, Knotkova Z, Hrda A, Dorrestein GM. Plasma bile acids in reptiles. *Proc Annu Conf Assoc Reptilian Amphib Vet.* 2009;8-15.

8. Knotkova Z, Dorrestein GM, Jekl V, et al. Fasting and postprandial serum bile acid concentrations in 10 healthy female red-eared terrapins (*Trachemys scripta elegans*). *Vet Rec.* 2008;163(17):510-514.

9. Reese SA, Jackson DC, Ultsch GR. Hibernation in freshwater turtles: softshell turtles (*Apalone spinifera*) are the most intolerant of anoxia among North American species. *J Comp Physiol B.* 2003;173(3):263-268.

10. Jackson DC. Hibernating without oxygen: physiological adaptations of the painted turtle. *J Physiol.* 2002;543(3):731-737.

11. Jackson DC, Taylor SE, Asare VS, et al. Comparative shell buffering properties correlate with anoxia tolerance in freshwater turtles. *Am J Physiol Regul Integr Comp Physiol.* 2007;292(2): R1008-1015.

12. Klein K, Adamovicz L, Phillips CA, et al. Blood lactate concentrations in eastern box turtles (*Terrapene carolina carolina*) following capture by a canine search team. *J Zoo Wildl Med.* 2021;52(1):259-267.

13. Innis CJ, Tlusty M, Merigo C, et al. Metabolic and respiratory status of cold-stunned Kemp's Ridley sea turtles (*Lepidochelys kempii*). *J Comp Physiol B.* 2007;177(6):623-630.

14. Withers PC, Lea M, Solberg TC, et al. Metabolic fates of lactate during recovery from activity in an anuran amphibian, *Bufo americanus*. *J Exp Zool.* 1988;246(3):236-243.

15. dos Santos BD, Alves AX, dos Santos NN, et al. Transport stress in bullfrog: hematological and plasma biochemical responses. *Aquac Rep.* 2021;19:100583.

16. Fournier PA, Guderley HE. Metabolic fate of lactate after vigorous activity in the leopard frog, *Rana pipiens*. *Am J Physiol Reg Integr Comp Physiol.* 1992;262(2):R245-254.

17. Gleeson TT, Bennett AF. Acid-base imbalance in lizards during activity and recovery. *J Exp Biol.* 1982;98(1):439-453.

18. Cerreta AJ, Cannizzo SA, Smith DC, et al. Venous hematology, biochemistry, and blood gas analysis of free-ranging Eastern copperheads (*Agkistrodon contortrix*) and Eastern ratsnakes (*Pantherophis alleghaniensis*). *PLoS ONE.* 2020;15(2):e0229102.

19. Arguedas R, Steinberg D, Lewbart GA, et al. Haematology and biochemistry of the San Cristóbal lava lizard (*Microlophus bivittatus*). *Conserv Physiol.* 2018;6(1):coy046.

20. Pfitzer S, Ganswindt A, Fosgate GT, et al. Capture of farmed Nile crocodiles (*Crocodylus niloticus*): comparison of physiological parameters after manual capture and after capture with electrical stunning. *Vet Rec.* 2014;175(12):304.

21. Lewbart GA, Hirschfeld M, Brothers JR, et al. Blood gases, biochemistry and haematology of Galápagos marine iguanas (*Amblyrhynchus cristatus*). *Conserv Physiol.* 2015;3(1)cov034.

22. Lewbart GA, Hirschfeld M, Denkinger J. et al. Blood gases, biochemistry, and hematology of Galapagos green turtles (*Chelonia mydas*). *PLoS ONE.* 2014;9(5):e96487.

The Fungi in the Room: *Nannizziopsis*

Krista Keller, DVM, Dipl ACZM

Session #2623

Affiliation: From Wildlife Epidemiology Laboratory, College of Veterinary Medicine, University of Illinois at Champaign-Urbana, Urbana, IL 61801, USA.

Nomenclature

An emerging fungal pathogen was increasingly diagnosed in reptiles in the late 1990's and early 2000's.[1] Based upon morphologic colony similarities noted from culture and the lack of a telomorph (sexual) phase able to be identified, all of these isolates were referred to as CANV (*Chrysosporium* anamorph of *Nannizziopsis vriesii*).[1] CANV was diagnosed and reported in a wide variety of reptilian species at that time, spanning most reptilian taxa and a strong causal relationship between the fungal agent and cutaneous and sometimes systemic mycosis was suspected.[1] In 2013, two separate studies evaluated available DNA sequences and determined that all isolates should be placed in three separate genera, *Nannizziopsis*, *Paranannizziopsis* and *Ophidiomyces*. *Paranannizziopsis* infection has been reported in tuatara and is the most important cause of morbidity and mortality in spectacled snakes under human care in North America and a single species in the genus *Ophidiomyces* (*O. ophidiolcola*) is the causative agent of snake fungal disease, a conservation concern in free ranging snakes.[2,3]

Nannizziopsis spp. are keratinophillic fungi, within the order Onygenales.[1] Infection has been documented in a variety of free ranging, companion and institutionally-housed lizard species (Table 1) as well as snakes, turtles, and crocodiles.[1-4] Recently, Koch's postulates have been proven for *N. guarroi* as the causative agent of yellow fungal disease in bearded dragons;[5] however, a variety of *Nannizziopsis* species has been associated with similar clinical signs in this species.[1-4] In Australia, the emergence of *N. barbata* has been implicated as a significant conservation concern for native lizard species.[6]

Table 1. Lizard species with reported *Nannizziopsis* spp. fungal infections.

Lizard species	Fungal species
Bearded dragon (*Pogona vitticeps*)	*N. chlamydospora*
	N. draconii
	N. guarroi
	N. barbata
Thai/Chinese water dragon	*N. arthrosporoides*
Skink (unknown species)	*N. pluriseptica*
Ameiva spp.	*N. vriesii*
Green iguana (*Iguana iguana*)	*N. guarroi*
Common agama (*Agama agama*)	*N. guarroi*
Chameleons	*N. dermatiditis*
Geckos	*N. dermatiditis*

There are currently no data supporting the incidence or prevalence of these infections in companion lizards; however, it is generally accepted that *N. guarroi* is the most common cause of cutaneous mycosis in companion lizards.[1,4]

Clinical Signs of Infection

Clinical signs range from mild scale discoloration to severe, disfiguring and ulcerative lesions, which can ultimately lead to invasive systemic mycosis and death.[1-4] In companion reptiles, infection with *Nannizziopsis* spp. typically begins as hyperkeratotic superficial cutaneous lesion; in the bearded dragon, these crusty skin lesion often have a yellow discoloration, leading to the colloquial term yellow fungus disease;[1,4] however, the lack of coloration should not exclude a fungal differential. Left undiagnosed or untreated, the cutaneous lesions invade into deeper tissues (muscle, bone) and can spread systemically to the liver, heart, and lungs.[1] This infection is associated with high morbidity from the primary mycotic infection and secondary bacterial invaders and mortality if not treated.[1,4]

Outside of the primary skin lesions, lizards often present underconditioned, dehydrated, lethargic and weak and may be suffering from comorbidities (e.g., nutritional secondary hyperparathyroidism, endoparasitisms).

Available Diagnostics

Diagnostics assessing systemic health can be helpful in detecting severity of underlying inflammation, dehydration and assessing organ health. Baseline diagnostics may include complete blood count, plasma biochemistry, whole body radiographs and fecal diagnostics to assess endoparasitism.

Diagnostics that could aid the practitioner in determining if a fungal pathogen was present could include:

Cytology

Crust cytology, stained with Diffquik can identify inflammatory cells, presence of bacteria and the presence of fungal elements. *Nannizziopsis* sp. Present cytologically as oval to cylindrical aleuoconidia, the asexual phase. More rarely hyphae will be seen. Although a mycologist may be able to differentiate the many species of *Nannizziopsis* based upon cytologic differences, this is rarely something that is able to be performed in practice.

Histopathology

A biopsy of the lesion can be sent in for histopathologic evaluation. Histopathology, with special stains (Grocott/methanmine silver or Periodic acid-Schiff) can confirm the presence of fungal hyphae that are suggestive of the genus *Nannizziopsis*. The histopathologic appearance of fungal agents in this genus are not specific at the species level.

Fungal culture

If a fungal agent is suspected or confirmed cytologically, a swab of the cutaneous lesion can be taken and sent for fungal culture. Because many *Nannizziopsis* spp have different incubation needs for growth, clinicians should consider sending samples to a laboratory with experience working with reptile fungal pathogens. Because all species within the genus have similar colony appearances, fungal culture itself cannot be used to determine which

Nannizziopsis species is present. Benefits of culture are the ability to perform antifungal susceptibility testing on the isolate, however, longer incubation times are required.

Diagnostics that can aid the practitioner in determining which *Nannizziopsis* species is causing infection should be pursued to allow for choice of empirical therapy based upon the available literature and more esoterically to build the lacking knowledge base surrounding the spread of this infection. These diagnostics include:

Panfungal PCR

A panfungal PCR will amplify a highly reserved region of the fungal genome, so any fungal DNA in the sample will be able to be amplified. The assay generally requires that the amplified DNA undergoes DNA sequencing to be compared to known sequences to determine the genus and species.

N. guarroi qPCR

A single laboratory (Wildlife Epidemiology Laboratory at the University of Illinois) has a quantitative PCR assay that is sensitive enough to detect small quantities of N. guarroi DNA (as few as 10 DNA copies per sample). The quantitative component of the assay can make evaluation of treatment success possible.

Matrix-assisted laser desorption/ionization time of flight mass spectrometry (MALDI-TOF MS)

This technology is increasingly used to rapidly identify species from cultures in laboratories. A single report has shown that MALDI-TOF was able to positively identify an isolate of *N. guarroi* from culture and this may represent a promising diagnostic in the future as additional species of *Nannizziopsis* are characterized.

Therapy

The sick lizard will greatly benefit from a holistic approach to improvement of health to include rehydration, administration of anti-inflammatories, assist feeding, correction of husbandry deficits, and treatment of concomitant bacterial infection at the site of skin ulceration.

The antifungal susceptibility of four clinically derived *N. guarroi* isolates have been reported in the literature and can be used to empirically choose an antifungal agent.[7] The same report explored the single oral dose pharmacokinetic profile of terbinafine in eight healthy adult bearded dragons. Each dragon was administered a single oral dose of terbinafine at 20 mg/kg and plasma sampled at 0, 4, 8, 12, 24, 48, 72, and 96 hours after administration. Plasma terbinafine levels were measured using a species validated HPLC. A peak plasma terbinafine concentration (Cmax) of 435 ± 338 ng/ml was measured at (Tmax) 13 ± 4.66 hours after administration. Plasma concentrations remained > 30 ng/mL (target plasma concentration based upon MIC data) for > 24 hours (8/8 bearded dragons) and > 48 hours (6/8 bearded dragons). Based upon these data and calculated accumulation ratios, a terbinafine dose of 20 mg/kg every 24-48 hours may be recommended.[7] Although multidose and safety studies have not been performed, the author routinely uses terbinafine 20 mg/kg by mouth once daily to successfully clear infection without any adverse effects.

Prior literature has evaluated two azoles, itraconazole and voriconazole in naturally infected bearded dragons.[8] Because the report predates 2013, the infection was reported to be caused by CANV. In this report, high mortality was noted in itraconazole treated animals (4/7) compared to much lower mortality reported in voriconazole treatment animals (1/7). This report and several other case reports support that itraconazole is a poor choice for the empirical therapy of infected bearded dragons.[8]

Environmental Cleaning

Current research supports that *N. guarroi* can be cultured from many surfaces (stainless steel, cotton, repticarpet) for at least 3 days after contact and from at least 14 days from aqueous environments.[9] This environmental persistence has major implications, as fomites may play an important role in the epidemiology of infection. The clinician should consider this when a suspected or confirmed infection status lizard has frequented the veterinary clinic to reduce the chance of transmission in the veterinary hospital. In addition, given the high rate of infection recrudescence reported in the literature, at home disinfection should be recommended. The environmental persistence of other species of *Nannizziopsis* is unknown, but expected to be similar.

A single study has evaluated effective disinfectants against *N. guarroi* and found that a 2% dilution of commercially available bleach with a 10-minute contact time or a 10% dilution of commercially available bleach with a 2-minute contact time is the only proven way to disinfect. This study showed that commonly used veterinary disinfectants including Virkon-S® and chlorhexidine (2%) did not have disinfecting capability.[10] Clinicians concerned with environmental contamination with *N. guarroi* in the veterinary office should ensure they are using appropriate protocols to reduce transmission between patients.

Prognosis

Unfortunately, the literature is filled with case reports of lizards not being able to clear infection, suffering from presumed antifungal toxicity, or infection recrudescence. There is currently a lack of data supporting which treatment protocol is the most appropriate to treat *Nannizziopsis* infected lizards. The author commonly treats *N. guarroi* infected bearded dragons and has had success treating with terbinafine 20 mg/kg PO q24, weekly environmental decontamination performed by the client, and tracking treatment progress with monthly *N. guarroi* qPCRs. Expected treatment courses should be expected to be 3-9 months.[11]

Human Health Significance

There are several reports of human infection with various *Nannizziopsis* spp. The reports appear to be limited to individuals with underlying health concerns that would be associated with immunosuppression. Osteomyelitis and abscessation of multiple organs is reported. Although a review in 2013 showed that human-associated isolates available for evaluation at that time were different than reptile-associated isolates, clinicians should be aware of the potential health concerns when diagnosing or suspecting *Nannizziopsis* infection in reptile patients and discuss the potential health concerns with those in contact.

References

1. Pare J, Sigler L. An overview of reptile fungal pathogens in the genera *Nannizziopsis*, *Paranannizziopsis*, and *Ophidiomyces*. *J Herpetil Med Surg*. 2016;26(1-2):46-53.

2. Sigler L, Hambleton S, Paré JA. Molecular characterization of reptile pathogens currently known as members of the *Chrysosporium* anamorph of *Nannizziopsis vriesii* complex and relationship with some human-associated isolates. *J Clin Microbiol*. 2013;51(10):3338-3357. doi: 10.1128/JCM.01465-13.

3. Stchigel AM, Sutton DA, Cano-Lira JF, et al. Phylogeny of chrysosporia infecting reptiles: proposal of the new family Nannizziopsiaceae and five new species. *Persoonia*. 2013;31:86-100.

4. Cabañes FJ, Sutton DA, Guarro J. *Chrysosporium*-related fungi and reptiles: A fatal attraction. *PLoS Pathog*. 2014;10(10):e1004367. doi: 10.1371/journal.ppat.1004367.

5. Gentry Sl, Lorch JM, Lankton, et al. Koch's Postulates: confirming *Nannizziopsis guarroi* as the cause of yellow fungal disease in *Pogona vitticeps*. *Mycologia*. 2021;113(6):1253-1263. https://doi.org/10.5818/1529-9651-26.1-2.46.

6. Peterson NR, Rose K, Shaw S et al. Cross-continental emergence of *Nannizziopsis barbatae* disease may threaten wild Australian lizards. *Sci Rep*. 2020;10;20976.

7. McEntire M, Reinhart JM, Cox SK et al. Single-dose pharmacokinetics of orally administered terbinafine in bearded dragons (*Pogona vitticeps*) and the antifungal susceptibility patterns of *Nannizziopsis guarroi*. *Am J Vet Res*. 2021;1-8.

8. Van Waeyenberghe L, Baert K, Pasmans F, et al. Voriconazole, a safe alternative for treating infections caused by the *Chrysosporium* anamorph of *Nannizziopsis vriesii* in bearded dragons (*Pogona vitticeps*). *Med Mycol*. 2010;48(6):880-885.

9. Durrante K, Foltin E, Cerreta A, et al. Environmental persistence of *N. guarroi*. *Proc ExoticsCon*. 2021;447.

10. Hemby C, Keller KA, Guzman DS, et al. Effectiveness of common disinfecting agents against isolates of *Nannizziopsis guarroi*. *Proc ExoticsCon*. 2019;567.

11. Foltin E, Keller KA. Successful treatment of *Nannizziopsis guarroi* infection using systemic terbinafine in the central bearded dragon, *Pogona vitticeps*. *J Herpetol Med Surg*. 2022;32(1);20-25.

The Fungi in the Room: *Emydomyces testavorans*

Krista Keller, DVM, Dipl ACZM

Session #2624

Affiliation: From Wildlife Epidemiology Laboratory, College of Veterinary Medicine, University of Illinois at Champaign-Urbana, Urbana, IL, 61801 USA.

Nomenclature

In the late 1990's and early 2000's an emerging fungal pathogen was increasingly diagnosed in squamates, tuatara and crocodilians.[1] Based upon similarities in morphology of colonies noted from culture and the lack of a telomorph (sexual) phase able to be identified, all of these isolates were referred to as CANV (*Chrysosporium* anamorph of *Nannizziopsis vriesii*).[1] DNA sequence analysis has determined that CANV isolates should be placed in three genera: *Nannizziopsis*, *Paranannizziopsis* and *Ophidiomyces*.[1-4] Although these genera represent important primary reptile pathogens that have individual and population-based effects, none of the pathogens have been noted to cause infection in aquatic chelonians.

In 2019, *Emydomyces testavorans* was first described after isolation from shell lesions in freshwater chelonians.[5] Molecular sequencing placed it into the Order Onygenales and this hallmark manuscript described the mycosis from a wide species breadth and from both animals under human care and free-ranging.[5] Given these findings, it is generally accepted that all aquatic chelonians are susceptible to infection although continued research in free-ranging populations and companion turtles will give more information on prevalence and incidence of infection.

Clinical Signs of Infection

External clinical signs are often associated with shell abnormalities and may include ulcerations of scutes, pitted depressions, discoloration and/or flaking of scute keratin.[6] Many cases have hard expansile nodular masses within the bone that are not detectable externally and are able to compress and distort internal coelomic contents.[6] When these nodular masses are explored, they often contain waxy dark colored debris and histopathologically represent epithelial inclusion cysts.[6] Given the range from mild to severe shell changes, some turtles may be completely subclinical while others show malaise, lack of appetite or other systemic effects.

In alligator snapping turtles (*Macrochelys temminckii*), infection with *E. testavorans* has a more cutaneous presentation, associated with cutaneous ulceration, hyperkeratosis, and loss of nails from distal limbs.[7]

Available Diagnostics

Diagnostics assessing systemic health can be helpful in detecting severity of underlying inflammation, dehydration and assessing organ health. Baseline diagnostics may include complete blood count, plasma biochemistry, whole body radiographs and fecal diagnostics to assess endoparasitism.

Diagnostics that could aid the practitioner in determining if a fungal pathogen was present could include:

Cytology

Cytologic scrapes, stained with Diffquik can identify inflammatory cells, presence of bacteria and the presence of fungal elements. *Emydomyces* sp. Present cytologically as oval to barrel shaped athroconidia, the asexual phase. More rarely hyphae may be seen. The presence of athroconidia may be very suggestive of infection with *Emydomyces testavorans*, however, other onygenalean fungi may have similar cytologic morphology.[5]

Histopathology

A biopsy of the shell lesion can be sent in for histopathologic evaluation. Histopathology, with special stains can confirm the presence of fungal hyphae that are suggestive of *E. testavorans*. Clinicians can request Grocott/methanmine silver or Periodic acid-Schiff to stain for fungal agents. The histopathologic appearance of fungal hyphae may be very suggestive of infection *E. testavorans* but are not specific at the species level.[5,6]

Fungal culture

If a fungal agent is suspected or confirmed cytologically, a deep swab of the shell lesion can be taken and sent for fungal culture. Because many reptile fungal pathogens have different incubation needs, clinicians should consider sending samples to a laboratory with experience working with these fungi. Because most onygenalean fungi have similar colony appearance, fungal culture itself cannot be used to determine if *Emydomyces testavorans* is present and downstream diagnostics from the fungal culture are required. Benefits of culture are the ability to perform antifungal susceptibility testing on the isolate. However, culture of this fungi can be very challenging requiring specialized media and prolonged incubation times.[5,6]

Diagnostic imaging

Radiographs and computed tomography (CT) may help detect cystic bone lesions. Given the superior detection of CT, radiographs should be considered as able to screen for inclusion cysts, while CT may be more definitive. Although the presence of bone cysts should be highly suggestive of infection *E. testavorans* they are not specific for infection as other infections may cause similar lesions.[6]

Diagnostics that can aid the practitioner in determining if *Emydomyces testavorans* is causing infection should be pursued to allow for choice of empirical therapy based upon the available literature and more esoterically to build the lacking knowledge base surrounding the spread of this infection. These diagnostics include:

Panfungal PCR

A panfungal PCR will amplify a highly reserved region of the fungal genome, so any fungal DNA in the sample will be able to be amplified. The assay generally requires that the amplified DNA undergoes DNA sequencing to be compared to known sequences to determine the genus and species.

E. testavorans qPCR

A single laboratory (ZooPath Laboratory at the University of Illinois) has a quantitative PCR assay that is sensitive enough to detect small quantities of *E. testavorans* DNA. Samples can be obtained antemortem from a shell swab, however swabs from deeper sites are more likely to yield a positive result. Postmortem samples should be obtained prior to placement of tissues into formalin and should focus on suggestive lesions of the shell.

Therapy

A clinically sick turtle will greatly benefit from a holistic approach to improvement of health to include rehydration, administration of anti-inflammatories, assist feeding, correction of husbandry deficits, and treatment of concomitant bacterial infection at the site of shell ulceration.

The antifungal susceptibility of several molecularly confirmed isolates of *E. testavorans* have been reported in the literature and can be used to empirically chose an antifungal agent.[5,8] Based upon available pharmacokinetic literature, voriconazole administered at 10 mg/kg SC q48h would exceed inhibitory concentrations in the Western pond turtle (*Actinemys marmorata*); however, there may be risk of injection site cutaneous sloughing.[8]

Terbinafine has also been evaluated in the published literature, and a single published pharmacokinetic study found that terbinafine administered orally (15 mg/kg) may require q12h administration to meet inhibitory concentrations for *E. testavorans*.[9] Unpublished research (Keller personal comm.) has shown that terbinafine 30 mg/kg may be able to be given every 24-48 hours to meet the MIC of *E. testavorans* in the Western pond turtle. Nebulization of western pond turtles with terbinafine has been shown to improve clinical lesions and maintain MICs against *E. testavorans* in the scute keratins for an extended period.[11]

Prognosis

At this time, there are no published reports of successful treatment or rate of progression after infection is confirmed. Given that this pathogen has only recently been described, ongoing research at multiple institutions will likely elucidate more information in the coming years.

Conservation Concerns

There are several endangered aquatic chelonian species that have been shown to be susceptible to infection. Further, *E. testavorans* has been implicated as a cause of death in many head started Western pond turtles and wood turtles (*Glyptemys insculpta*).[11,12]

References

1. Pare J, Sigler L. An overview of reptile fungal pathogens in the genera *Nannizziopsis*, *Paranannizziopsis*, and *Ophidiomyces*. *J Herpetol Med Surg*. 2016;26(1-2):46-53.

2. Sigler L, Hambleton S, Paré JA. Molecular characterization of reptile pathogens currently known as members of the *Chrysosporium* anamorph of *Nannizziopsis vriesii* complex and relationship with some human-associated isolates. *J Clin Microbiol*. 2013;51(10):3338-3357. doi: 10.1128/JCM.01465-13.

3. Stchigel AM, Sutton DA, Cano-Lira JF, et al. Phylogeny of chrysosporia infecting reptiles: proposal of the new family Nannizziopsiaceae and five new species. *Persoonia*. 2013;31:86-100.

4. Cabañes FJ, Sutton DA, Guarro J. Chrysosporium-related fungi and reptiles: A fatal attraction. *PLoS Pathog*. 2014;10(10):e1004367. doi: 10.1371/journal.ppat.1004367.

5. Woodburn DB, Miller AN, Allender MC, et al. *Emydomyces testavorans*, a new genus and species of onygenalean fungus isolated from shell lesions of freshwater turtles. *J Clin Microbiol.* 2019;57(2):1-11.

6. Woodburn DB, Kinsel NJ, Poll CP, et al. Shell lesions associated with *Emydomyces testavorans* infection in freshwater aquatic turtles. *Vet Pathol.* 2021;58(3):578-586. doi.org/10.1177/0300985820985217.

7. Adamovicz L, Woodburn DB, Boers K, et al. Characterizing the clinical course of an emerging fungal pathogen (*Emydomyces testavorans*) in the state-endangered alligator snapping turtle (*Macrochelys temminckii*). *Proc Annu Conf Am Assoc Zoo Vet.* 2019.

8. Wright TL, Gjeltema J, Wack RF, et al. Plasma voriconazole concentrations following single and multiple-dose subcutaneous injections in Western pond turtles (*Actinemys marmorata*). *J Zoo Wildl Med.* 2021;52(2):538-547.

9. Eshar D, KuKanich B, Avni-Magen N, Joo H. Terbinafine pharmacokinetics following single-dose oral administration in red-eared slider turtles (*Trachemys scripta elegans*): A pilot study. *J Zoo Wildl Med.* 2021;52(2):520-528.

10. Flaminio KP, Cox S, Haman K, et al. Pharmacokinetics of nebulized terbinafine in the plasma and keratin of Northwestern pond turtles (*Actinemys marmorata*) associated with emydomycosis. *J Herpetol Med Surg.* 2022;32(1):48-55.

11. Browning G, Wright L, Wack RF, Gjeltema J. Fungal shell disease associated with *Emydomyces testavorans* in a zoo collection of western pond turtles (*Actinemys marmorata*). *Proc Annu Conf Am Assoc Zoo Vet.* 2019.

12. Milnes E, Dutton C, Laroughe C, et al. An epizootic of mycotic shell disease associated with Nannizziopsiaceae fungal infection in juvenile wood turtles (*Glyptemys insulpta*) in a conservation head-starting program. *Proc Annu Conf Am Assoc Zoo Vet.* 2018.

Gross Lesion Recognition for the Reptile Clinician

Michael M. Garner, DVM, DACVP

Session #2566

Affiliation: From Northwest ZooPath, 654 West Main Street, Monroe, WA 98272, USA.

Abstract: Gross lesion recognition, like other imaging modalities, is a bit of an art form. Vast zones of grey may confront the investigator when considering what the lesion is, what it could be, and what to do with it. And nobody is good at all of it. A seasoned livestock pathologist could possibly clear an entire day's necropsies without the help of a single histology slide. A bee keeper would recognize immediately the maggot stages that parasitize his colony. A dog-cat pathologist in a busy private laboratory or a companion animal clinician may not have a clue about gross lesions in livestock or bees. Historically, our attempts at accurately diagnosing disease by gross lesion recognition have been a humbling experience, and that has been the impetus for all further diagnostic specialties. So now that we have put gross pathology in its proper place, why even bother with such an imperfect science? Obviously, gross lesions, or lack thereof, are the first visual indication of what may be wrong with the patient. These are the lumps, the effusions, the asymmetrical oddities, the discolorations, the odiferous clues to disease that (hopefully) stimulate a scientific thought process culminating in a list of differential diagnoses. The purpose of this master class is to present images of common and not-so-common reptile gross lesions in live and necropsy specimens, and in a participatory manner, establish a differential diagnosis and means for establishing a definitive diagnosis.

Recommended Reading

1. Campbell TW, Ellis C. *Avian and Exotic Animal Hematology and Cytology.* Ames, IA: Iowa State University Press; 2007.

2. Garner MM. An overview of biopsy, cytology and necropsy techniques for clinical veterinarians. In: Aguilar RF, Hernandez SM, Divers SJ, Perpinan D, eds. *Atlas de Medicina de Animales Exoticos.* 2 da Buenos Aires: Inter-Medica S.A. I.C.I.; 2010:437-443.

3. Garner MM. Overview of biopsy and necropsy techniques. In: Mader D, ed. *Reptile Medicine and Surgery.* 2nd Ed. St. Louis, MO: Elsevier Saunders; 2006;569-580.

4. Stacy BA. Reptile necropsy techniques. In: Jacobson ER, Garner MG, eds. *Reptile Diseases and Pathology.* Vol 1, 2nd Ed. London, UK: CRC Press; 2021:331-374.

The Eyes of Reptiles and Amphibians: Forms and Functions

Jeanette Wyneken, PhD

Session #2617

Affiliation: From Florida Atlantic University, Boca Raton, FL 33431-0991, USA.

Abstract: Reptilian and amphibian eye anatomy grossly resembles that of other vertebrates. Aquatic anurans, salamanders and larval forms lack eyelids; terrestrial adults have reduced eyelids. All reptiles have eyelids; however, some lizards have partially fused lids and other lizard species, such as lidless geckos, and all snakes have fused clear lids (spectacles). Orbital glands lubricate the cornea in all species. The globe is formed of layers (sclera, choroid, and retina) and has three chambers (anterior chamber, posterior chamber, and vitreous chamber). The lens is soft and is shaped for accommodation by the ciliary muscles in amphibians, lizards, chelonians, and crocodilians or by lens movement (snakes). Specializations of the retinas' receptor cell forms, their organization in the retina, presence or absence of oil droplets in the cones, and eye speed (flicker fusion frequencies) reflect adaptations to the natural environments in which the animals live. The retina's photosensitive cells (rods and cones) transmit signals to the optic nerve, which is an extension of the brain. Retinal acuity is increased in foveae in lizards and tuataras, while amphibians, turtles and snakes lack foveae but may have visual streaks or specialized areas of concentrated cones. Color vision is documented in most species tested and recently sensitivity to UV wavelengths has been found in some amphibians and reptiles.

Introduction

This overview of amphibian and reptilian eyes covers the three major topics: normal gross anatomy of amphibian and reptilian eyes and ocular adnexa; eye function; amphibian and reptile use of vision. The anatomy of the eye is similar across species generally, but the eyes of various taxa differ in their details. Here anatomic and functional aspects are described in overview then greater detail is provided in the latter parts of this review. Amphibian eyes are adapted for aquatic vision in larval life stages and transition in form during metamorphosis for aerial or aerial and aquatic vision.[1,2] Anuran eye size significantly correlates with habitat, such that many terrestrial species have extremely large eyes and yes are much smaller in fossorial, sub-fossorial, and aquatic species.[3] Caecilian eyes are unusual and are summarized at the end of this review. Reptile eyes may be adapted for either aerial or aquatic vision depending upon species and habitat; unlike amphibian eyes which may undergo ontogenetic shifts.[4]

Gross Anatomy

The globe (i.e., eyeball) is formed of layers, surrounding a viscous medium, and is structured as a series of chambers (Figure 1). The globe's outer layer, the sclera, is a fibrous outer capsule to which extrinsic eye muscles attach) and in some taxa a ring of boney scleral ossicles gives the eye stability.[4-6] The sclera is usually white but can be strikingly pigmented. For example, the Cayman blue iguana (*Cyclura lewisi*) has a red sclera.[4]

The ocular adnexa include, at least, the extraocular muscles, eyelids, and orbital glands. Some references include the orbits and optic nerves.

Orbits

The body contributions to the orbits varies with major taxon. In frogs, major reductions and fusions result in orbits bounded by the maxilla, nasal, frontoparietal, and pterygoid bones. Salamanders have separate frontal and parietal bones and a prefrontal. The two amphibian eyes lack substantial intracranial separation and the ventral surface of each globe is adjacent to the tissue lining the roof of the mouth. Consequently, the eyes are retracted int the head during prey handling and swallowing.

In reptiles, other than snakes, the orbits are bounded by the maxilla, jugal (in tuatara, chelonians, and some lizards), postorbital, frontal, and prefrontal bones. The floor of the orbits is formed by the lacrimal, palatine, and sometimes the vomer. Snake skulls have reduced skeletal elements; the orbits are bordered by the prefrontal, supraorbital, and postorbital bones. The orbits are separated from one another by a cartilaginous interorbital septum in lizards, crocodilians and turtles.[1,4,11] Snake interorbital septa form from boney processes from frontal, parietal, and parasphenoid bones, often with an interorbital membrane. The orbital bones and the septum are lined with a periorbital membrane, which joins with the orbital membranes, and fuses with the proximal internal parts of the upper and lower eyelids. The periorbital membrane is traversed by veins and venous sinuses.[39]

Layers

The middle layers of the eye are collectively termed the uvea and include the choroid, tapetum lucidum (when present), ciliary body, and iris. The choroid is a vascular nutritive layer in most, but not all taxa. The ciliary body located just behind the iris has zonular fibers that attach to the lens; it is in the anterior portion of the uvea. Its major functions are production of the fluid in the anterior chamber and, in species that accommodate via lens shape, it helps shape the lens.[4,5] The iris and the pupil are the eye's aperture (constricted = miosis; dilated = mydriasis), controlling the amount of available light reaching the lens and retina[1,4,5,6] and may aid in accommodation in some chelonians.[6,8] The literature describing amphibian pupil aperture control is conflicting regarding muscle types (smooth, striated, or both).[1,11,12] The pupil size is controlled by striated muscles in reptiles.[1,6,9] This contrasts with smooth muscles controlling the iris in most vertebrates. The lens is supported by uveal structures.

The inner layer of the eye is the retina, which itself is organized into layers. The photosensitive cells (rods and cones) are inner-most. Light passes through a layer of ganglion and Amacrine cells, then a layer of interneurons, and then reaches the photosensitive cells. The optic nerve transmits the retinal signals to the brain. It exits the eye at the optic disk.[1,4,5,8,13]

The innermost wall of the globe is supported by hyaline cartilage in some amphibians (anurans) and many reptiles. Sometimes calcified cartilage is present in turtles, lizards, and crocodilians. Snakes lack cartilage in the sclera. A ring of multiple small bony plates, the scleral ossicles, (Figure 1) is located where scleral and corneal tissues join in supports the outer portion of the eye in lizards, turtles, and tuataras. [1,4,5,9,14-16] Amphibians, snakes and crocodilians lack scleral ossicles.[4,6,9,14,16]

The chambers of the eye are the outer chamber, defined as the space between the cornea and the iris and contains aqueous humour; the posterior chamber is demarked by back of the iris, bounded by the ciliary body around its circumference, and anterior and around the lens up to the suspensory ligaments. The vitreous chamber is the large chamber between the lens and the retina containing the vitreous body.[4]

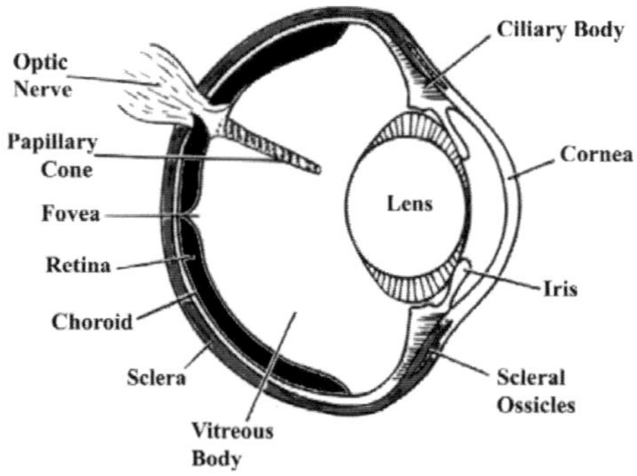

Figure 1. Generalized cross section of a reptile eye to show the major structures. Note that not all reptiles have a papillary cone. See Whitaker (2001) for an excellent diagram of an amphibian eye.[14]

Cornea

The cornea is clear, formed of thin layers, and is thin walled.[4,6,13] In many species, the cornea is tall and the lens is relatively flat on the side nearest the cornea. It functions as a lens and refracts light except underwater.[4,6,15] In aquatic reptiles, the cornea tends to be domed with a curvature that is similar to the sclera. In crocodilians, the cornea is flatter than might be expected for aquatic species and appears to do little refraction.[4] It is likely that other crocodilian sensory systems (such as dome receptors that detect waves) and hearing play more important roles than vision. Snakes too have a relatively flat cornea.[1,5,6]

Pupils

The pupil acts as the aperture setting for the eye. Pupil diameter (pupillary reflex) is controlled by the oculomotor nerve.[5,6] Pupil shape differs among within amphibian and reptilian taxa and with function and behavior.[4,12] The shape of the pupil can have a profound effect on the retinal image because pupil shape influences the characteristics of light reaching the retina.[1,6,15] Urodeles, turtles, diurnal lizards and colubrid snakes tend to have round or oval pupils.[4,12,18] Anurans have great diversity in pupil shapes. Aquatic species tend to have round pupils; ironically, diversity in pupil shape des not correlate well with habitats in other taxa.[12] Nocturnal hunters such as crocodilians, geckos, and many snakes have slits. Snakes may have elliptical slit or round pupils. The crocodilian pupil is a vertical ellipse that can contract to a slit. Slit pupils tend to improve focus in an orientation that is perpendicular to the slit.[4,5,9] Hence, a vertical pupil provides the best focus along the horizontal axis relative to the animal's head. Schwab[18] notes that it is important to remember that an animal's horizon may be different from your own. For example, geckos operate in diverse positions, often on vertical perches or on ceilings.

Horizontal pupils, common among herbivores, and anurans but rare or absent in reptiles. They focus best along the vertical axis. Slit pupils with multiple pinholes, found in some geckos and some anurans allows the animals to use the pinhole images as a range finder by forming several images of an object on the same point on the retina. If an object is too close or too far, the images from the individual pinholes will not be in focus or be coincident with one another.[18,19] Additionally, slit pupils with multiple apertures can dilate to a much larger field of view than a round pupil. Such dilation is advantageous for nocturnal activity.[18]

Intrinsic muscles

The amphibian pupillary sphincter is formed from smooth muscle, or possibly a combination of smooth and striated muscle.[1,4,11] In reptiles, striated muscle is responsible for pupillary size in reptiles, except for snakes. Snakes have smooth muscle in the iris controlling pupil size.[1] Thus, most reptilian eyes differ from the pupillary muscular system in mammalian eyes (purely smooth muscle sphincter). Striated ciliary muscles shape the lens directly or by pulling on or relaxing the zonular fibers.[1,4,5,11] The ciliary muscles may also push on the iris and squeeze part of the lens through the pupil, slightly reshaping it to focus. Snakes focus by moving the lens toward the iris or retina via pressure in the vitreous body from the iris.[20,22] Amphibians, some turtles, lizards, and possibly snakes may move the lens nasally and medially with the transversalis muscle for binocular vision.[4]

Retina

The photosensitive cells are overlaid by a layer of interneurons (horizontal and bipolar cells) and a layer of ganglion cells and Amacrine cells; the latter layer is adjacent to the vitreous chamber. The photosensitive cells of amphibians and reptiles are of two characteristic types, rods and cones. Typically rods detect light at low intensity levels but not color, except in anurans. Cones are active in bright light and typically are thought to be color detectors.[1,4,6,10] Cones occur in photoreceptors populations pigments that respond to different wavelengths. To provide for color vision, there must, minimally, be two different photopigments present and the neural organization to process the response differences.[4] In general, the ratio of rods to cones is related to the behavior of the animals, so that day active animals tend to have a higher proportion of cones and nocturnal animals have a higher proportion of rods; however, amphibians and reptiles have may specialization such that rods or cones may take on novel forms and functions. Using diverse behavioral assays, a variety of anuran species have been shown to perceive color and have unique rods (green rods) that detect color under low light conditions.[2,23]

Reptiles show considerable variation photoreceptors and functions, likely due to the many ecological shifts reptiles have undergone in their evolutionary history, many linages switched between diurnal and nocturnal life, and some shifted back). Consequently, some species have photoreceptors that express atypical forms or atypical photopigments.[6,15,18,23,24] Diurnal lizards and at least some diurnal snakes, such as garter snakes (*Thamnophis sirtalis*), have all-cone or nearly all-cone retinae.[4,6] Some nocturnal lizards, such as Tokay geckos (*Gecko gecko*) have retinae formed of only rods; however, crepuscular (dawn/dusk active) lizards and tuataras have all-cone retinae, yet the cones have secondarily developed rod-like function.[4,6,23] Nocturnal lizards and nocturnal or crepuscular snakes usually have rods and cones;[1,4,6,25] however, the rods may be secondarily derived from cones.[1,4,6] Turtles and crocodilians have retinas formed of both rods and cones.[1,4,6,10,15,26,30] Most, if not all, diurnal reptiles appear to see color. The numbers of cone photopigments and their spectral sensitivity provide presumptive evidence for the ability to detect colors;[17,26,27] however, detection (a physiologic response) and perception (the behavioral evidence that detected colors can be discriminated) may differ. The peak sensitivity to colors varies with species, and sometimes with age. Physiologic measures of the retina's sensitivity to colored lights gives indication of how many photopigments are present and the ranges of colors (wavelengths) they can detect;[4] however, whether the animals perceive the colors requires behavioral assays.[23,34] Many studies purporting to demonstrate color vision do not definitively do so because they fail to test for perception or control for light intensity differences among the wavelengths (colors) tested.

Eye speed

The visual temporal resolution or flicker fusion frequency (FFF) is the frequency at which oscillating movements or flickering light pulses that are detected as constant that than discrete events. Ecological drivers of the FFF include the movement of common prey or predators. In captive situations, eye speed may matter when artificial lighting is used. Animals that have FFF above 120 hz may detect may the 50-60 hz flickering of incandescent lamps or the 100-120 hz flicker of fluorescent bulbs.[33]

Photo receptor filters

Oil droplets may be found in the ends of cones.[4,35,36] They represent a common mechanism by which retinal sensitivity is modified.[32] They may be clear or colored and function as filters and can increase the ability of an animal to discriminate objects or edges and detect colors.[15,36] There is some evidence that oil droplets may protect the cones from UV damage.[4] Oil droplets are found in turtles and diurnal lizards, but are not reported in crocodilians or snakes.[1,36]

Foveae

Foveae are pits in the retina containing concentrated cones that provide for increased acuity. Foveae with steep walls tend to increase acuity by causing slight expansion of the image. in lizards and tuataras have foveae. Lizards have one or two foveae per retina; they are shallow in some species (e.g., iguanas and geckos) and deep in others, such as chameleons. Tuataras have a deep-wall fovea in each eye. Turtles (with the exception of *Amyda* spp., the soft-shelled turtles), crocodilians, some lizards, and snakes lack foveae.[1,4,10,27]

Areas and visual streaks

These are sections of the retina that are particularly sensitive due to increased number or density of photoreceptors or ganglion cells.[4,10,19] Several species of anurans have area or broad visual streaks.[28] Turtles have areas (e.g., an area centralis or area temporalis) or horizontal strips of concentrated photodetection called visual streaks, which are sections of the retina that are especially sensitive.[15,19,30] Crocodilians and several lizard taxa also have visual streaks.[4,19,29,31,32] Visual streaks have been identified in several species of sea snake, and in the garter snake.[4,24] (Snakes supplement vision with acute chemical sensitivity and, in some cases, infrared radiation detection.) Areas and visual streaks are linked closely with the visual ecology of the animal. These specializations of the retina have enhanced sensitivity and help the animals avoid predators, find essential habitat and find food.[4]

Optic disc and papillary cone

The optic disc is the region in the retina where nerve fibers from the retina exit to become part of the optic nerve. In some reptiles, a papillary cone (*conus papillaris*) arises from the optic disc and extends into the vitreous body.[1,4,11] It is often melanic and highly vascular. It is believed to provide supplementary nutrition to deep ocular tissues, replacing or supplementing the role of the choroid. A papillary cone is present in lizards and some snakes, but not in turtles, crocodilians or the tuatara. Some snakes supply nutrition to the retina via choroidal diffusion, while others rely upon a network of blood vessels lying on top of the retina (vitreal vessels).[4,10,11]

Chambers

The chambers of the eye are the anterior chamber, which is a fluid-filled space inside the eye between the iris and the cornea's innermost surface; the posterior chamber which is a small space directly posterior to the iris, anterior to the lens, and bordered by the ciliary body or ciliary muscles. The anterior and posterior chambers are filled with aqueous humour. The vitreous chamber is filled with a viscous liquid, the vitreous body (i.e., vitreous humour), within the chamber between the lens and the retina.[1,11,14]

Pupils

Pupil shape differs among taxa and often is linked behavior and ecology. Pupil shape and diversity is described for anurans, a few urodele, lizards, most chelonians, and some snakes. Very little research has been done on pupillary light reflexes in intact amphibians and reptiles and few species have been examined (three species of anurans, one urodele, and two species of reptiles).[4,12,13] The few amphibian species studied appear to have smooth muscle controlling the aperture. The pupillary sphincter in reptiles is formed from striated muscle or a combination of smooth and striated muscle (as in some birds like chickens). This arrangement differs from the purely smooth muscle sphincter found in mammals. In general, pupillary light responses are described as sluggish in amphibians and red-ear sliders (*Trachemys scripta*) and somewhat faster in the green iguana (*Iguana iguana*). Pupil diameter (pupillary reflex) is controlled by the oculomotor nerve (CN III). Systematic study of amphibian pupillary muscles has not been done. There is limited evidence for voluntary pupil control in freshwater turtles. Consensual pupil response has not been document in amphibians and is partial in red-ear sliders.[12,22,37]

Lens

The lens is soft and is shaped for accommodation by the ciliary muscles (lizards, turtles, and crocodilians) or movement (snakes). Amphibian lenses are round in larval forms but flatten during metamorphosis in taxa that become terrestrial. The ciliary body of anurans has anterior-posterior folds. Folds are absent in urodles.[7] Striated ciliary muscles may help shape the elastic lens and cornea in reptiles, except in snakes. These muscles are innervated by CN VI and, in some cases, CN VII.[11,14,21,38]

Color Vision or Not

The literature on photosensitive cells rods and cones is abundant. In very general terms, rods tend to detect light at low levels. Cones tend to detect colors and are active in bright light. Both amphibian and reptilian taxa violate this generality. Few amphibians have been examined and the visual pigments suggest a limited range of color vision in some.[2] Turtles have rods and cones. The tuatara has cone-like photoreceptors that have rod-like features. Crocodilians have retinas formed of both rods and cones (there are more rods).[1,4,17] Nocturnal lizards and snakes have rods and cones (the latter may be secondarily derived from rods).[6] Diurnal lizards have all-cone retinae.[11,19] Most diurnal reptiles can see color. The peak sensitivity to colors varies with species, and sometimes with age. Behavioral assays are necessary to determine whether color is detected (a neurophysiologic response) or perceived (behavioral responses).[34]

Eye Movements Via Extrinsic Eye Muscles

Eye movement is essential to prevent fatigue of photoreceptors.[4] Turtles, crocodilians, and most lizards (except *Heloderma* spp.) have mobile eyes. Snake eyes, effectively, are not mobile.[10,11,21] Reptiles have *retractor oculi* (i.e., retractor bulbi) and *protractor oculi* (i.e., levator bulbi) muscles inserting on the sclera adjacent to the optic nerve that move the eye inward and outward within the socket.[1,4,6,9] These muscles are weakly developed in crocodilians, some turtles and many lizards without hinged snouts and braincases (akinetic lizards). They are strongly developed in lizards with kinetic skulls, chameleons, most turtles, and snakes.[1,6,11] Cranial nerve VI (abducens), and possibly cranial nerve VII (facial), innervate(s) them. Amphibians and reptiles have six extrinsic eye muscles that are responsible for rotating the globe within the orbit (Table 1); they insert on the sclera.[5] The extrinsic eye muscles are characterized by fine movement control and coordinated control of eyes with the vestibular system.[4,5,14]

Table 1. Extrinsic eye muscles, their innervations and actions.[3,6]

Muscle	Innervation	Action[a]
Superior rectus	Cranial nerve III (Oculomotor)	Draws gaze temporally and dorsally
Inferior rectus	Cranial nerve III (Oculomotor)	Draws gaze nasally and ventrally
Medial rectus	Cranial nerve III (Oculomotor)	Draws gaze nasally
Inferior oblique	Cranial nerve III (Oculomotor)	Draws gaze temporally and ventrally
Superior oblique	Cranial nerve IV (Trochlear)	Draws gaze nasally and dorsally
Lateral rectus	Cranial nerve VI (Abducens)	Draws gaze temporally

[a]The innervations are conservative. Because of species-specific differences in orbit shape and size affect muscle insertion position, these actions are given in the most general terms.

Accommodation

Lenses and focusing mechanisms: Accommodation involves the lens and the cornea in terrestrial reptiles.[1,2,4,9] The cornea plays an important role in light refraction in terrestrial amphibians and reptiles. It plays a lesser role in refracting light in aquatic taxa.[2,4,9] The steeper the curvature of the lens or the cornea, the greater the light gathering capacity.[2,11] The lens is a composite of relatively soft crystalline proteins. In aquatic amphibian and reptile species, the lens tends to be round. In most semiterrestrial and terrestrial amphibians, the lens has a flattened shape and the eye is more spherical. The cornea is the major refractive structure in terrestrial taxa and accommodation occurs by moving the lens as a whole rather than shaping the lens shape ciliary muscles.[1,2,4,6] A *protractor lentis* muscle moves the lens toward or away from the retina in amphibians rather than changing its shape. Reptilian lenses are often oval and are focused by deformation (i.e., lenticular deformation). Lens deformations largely are due to lens elasticity, by the force of circular muscle fibers in the iris and relaxation of zonular fibers.[12] Aquatic reptilian lenses are round and deformable.[12] Snakes have round lenses that are moved toward or away from the retina to focus.[1,21]

Eye Shine or Not

Vision under low light conditions may be facilitated by a tapetum lucidum. The tapetum lucidum (reflective material) is in the choroid.[11,19] No tapetum is present in amphibians, (with the possible exception of a few toads), snakes, lizards, turtles, or the tuatara;[4] however, the tapetum of crocodilian is beneath and within the outer retina rather than in the choroid.[19]

Eyelids

All reptiles and terrestrial amphibians have eyelids. Fully aquatic anurans and all larval stages lack eyelids. Crocodilians, chelonians, many lizards, and the tuatara, have mobile eyelids. Most lizards that have movable lids have a cartilaginous plate, the tarsus, in the lower lid.[4,11] Some lizards have partially fused lids (i.e., chameleons). Other lizards (i.e., geckos, skinks) and all snakes have fused clear lids, the spectacle, which is highly vascularized. Between the fused lids and the cornea is the subspectacular space which is drained by the nasolacrimal duct.[10]

Amphibian eyelids and eye closing. In fully aquatic species and larval life stages, no eyelids are present or are very rudimentary (i.e., axolotls, African clawed frogs). Other urodele taxa (mostly aquatic and some semiaquatic species) have mobile lids. Terrestrial salamanders usually lack lids. Eyelids are present in frogs and toads after metamorphosis. The upper lid is immobile or moves little and the lower lid is elevated to close the eye.[14] The lower lid tends to be translucent in many species. Anurans tend have a third eyelid (i.e., nictitans) but none is present in urodeles.[2] The *retractor bulbi* muscle inserts on the ventral and medial aspect of the globe and to the rim of the lower lid via a tensed cord of tissue that pulls the globe down and inward. This action effectively pulls the eye inward and closes the lower lid.[4]

In reptiles, the upper lid has mostly smooth muscle and is less mobile than the lower lid with striated muscle. Both upper and lower eyelids are present in lizards (in most cases), all turtles, and crocodilians. The lower lid of lizards contains a cartilaginous support.[4] In crocodilians, the opposite is the case; the upper lid, containing a bony plate, closes the eye,[5] and the lower lid lacks bone or cartilage and moves little. The borders of the upper and lower lids are often rich in goblet cells (secretory)[1]. Reptilian eyelids cover a poorly cornified nictitating membrane along the nasal surface of the eye; the nictitating membrane is a cartilage-supported extension of the conjunctiva. Nictitating membranes are highly developed in crocodilians and turtles. The nictitating membrane, which may be pigmented, is usually largest toward the medial (nasal) part of the eye.[4,6] Depending upon species, it may cover all or part of the eye. Lizards generally lack nictitating membranes. Eyelids are modified in a number of species. Some skinks, lacertid, and iguanine lizards have a transparent lower eyelid. Chameleons have partially fused upper and lower lids and are reported to lack a nictitating membrane.[1,10,11] While most lizards have eyelids, most geckos and Australian snake lizards tend to lack them.[1,4,37,39] The cornea of snakes and many geckos is covered by a transparent spectacle, which is derived from fused eyelids. Snakes lack eyelids and nictitating membranes. and its outer layer is shed with the skin.[6,10,11]

Orbital glands

Orbital glands lubricate the cornea in all species.

Amphibian orbital glands protect eyes that are exposed to air rather than water. Lacrimal glands in urodeles (not anurans) lubricate the eye; they lie in the superior temporal region over the globe, in the anterior part of the orbit. Lacrimal secretions are mixed with mucus secreted by scattered cells in the conjunctiva. Most frogs have a Harderian gland located along the media; nasal aspect of the globe. The Harderian gland develops at metamorphosis.[4,14,15] The Harderian glands are hypertrophied and occupy the nasal half of the orbit in caecilians.[1] Some urodeles have a mobile lower lid that contains separate sets of secretory glands nasally and temporally.[15]

Reptilian orbital glands vary with taxa.

Lizards generally have three orbital glands which may be compact or have extended portions. Most lizards have well-developed lacrimal glands posterior, dorsal and ventral to the eye (absent in chameleons, calotes, some geckos, and Australian snake-lizards). Harderian glands anterior to each globe drain via a duct onto the inner surface of the nictitating membrane.[1,6,11] The Harderian ducts extend from the lower lid and empties into the palate. A small mucous gland (sometimes termed the conjunctival gland) opens onto the outer surface of the nictitating membrane. Snakes have well-developed Harderian glands, which lubricate the spaces between the spectacle and the cornea. A second duct drains this fluid into the Jacobson's organ in the dorsal region of the oral cavity. Snake Harderian glands are located dorsally and nasally. In turtles, lacrimal and Harderian glands vary in size with taxon. They are very large in marine turtles but small in freshwater species. There is a single lacrimal duct in turtles located posterior to the globe.[1,5,6] The Harderian gland is present dorsally and nasally in all turtles.[11] Crocodilians have three kinds of glands associated with the eye: lacrimal, Harderian, and conjunctival glands.[5,36] The lacrimal gland is an elongated gland that is small relative to the globe size and located dorsally within the orbit. The Harderian gland is large, triangular and located nasally and medial to the eye. This gland secretes lubricating fluid via two ducts that drain between the nictitating membrane and the eye. The conjunctival gland is located at the junction of the conjunctiva and the globe within the lower lid. It is presumed to be a lubricating gland as well.[9]

Caecilians (Amphibia, Gymnophiona)

Several frequently cited references report that caecilian eyes are vestigial; however, Wake[45] examined caecilians in some depth and concludes that the caecilian eye morphology includes the following trends. The eyes form and are overlain by thick, often glandular skin, or by bone as well as skin. The extrinsic eye muscles are reduced, and some to all may be lost. The retina is present and varies in its architecture with taxon such that some taxa have the typical vertebrate layered retinal organization, others have reduced cell number, and yet other taxa have a retina that is net-like rather than layered. The optic nerve is present in some form such that it is attenuate in some taxa and possibly represented only by glial cells in others. The lens, when present, is round (aquatic forms, larval and adult) to spheroid; it ranges from crystalline to cellular, to amorphous to absent. The lens is free to adherent to cornea to adherent to both cornea and retina. The vitreous body is reduced or lost. The cornea adheres to the overlying dermis or periosteum. The scolecomorphids have eyes that are displaced out of the sockets and embedded in the tentacle under the upper jaw skin.

References

1. Walls GL. *The Vertebrate Eye and Its Adaptive Radiation.* Bloomfield Hills, MI: Cranbrook Institute of Science; 1942.

2. Mohun SM, Davies WI. The evolution of amphibian photoreception. *Front Ecol Evol.* 2019;7:321.

3. Thomas KN, Gower DJ, Bell RC, et al. Eye size and investment in frogs and toads correlate with adult habitat, activity pattern and breeding ecology. *Proc R Soc Med B.* 2020;287:20201393.

4. Schwab IR. *Evolution's Witness: How Eyes Evolved.* New York, NY: Oxford University Press; 2012.

5. Kardong KV. *Vertebrates. Comparative Anatomy, Function and Evolution.* 8th Ed. New York, NY: McGraw-Hill Co; 2018.

6. Underwood G. The eye. Pp. In: Gans C, Parsons T, eds. *The Biology of the Reptilia.* Vol. 2. Morphology B. New York, NY: Academic Press; 1970:1-97.

7. Keller CB, Shilton CM. The amphibian eye. *Vet Clin North Am Exot Anim Pract.* 2002;5(2):261-274.

8. New ST, Bull CM. Retinal ganglion cell topography and visual acuity of the sleepy lizard (*Tiliqua rugosa*). *J Comp Physiol A Neuroethol Sens Neural Behav Physiol.* 2011;197(6):703-709.

9. Holmberg BJ. Ophthalmology of Exotic Pets. In: Maggs D, Miller P, Ofri R, eds. *Slatter's Fundamentals of Veterinary Ophthalmology.* Elsevier Saunders; 2008:427-441.

10. Jacobson E. *Infectious Diseases and Pathology of Reptiles: Color Atlas and Text.* Boca Raton, FL: CRC Press; 2007.

11. Lawton MPC. Reptilian ophthalmology. In: Divers SJ, Stahl SJ, eds. *Mader's Reptile and Amphibian Medicine and Surgery.* 3rd Ed. St. Louis, MO: Elsevier; 2019:721-735.

12. Douglas RH. The pupillary light responses of animals; a review of their distribution, dynamics, mechanisms, and functions. *Prog Retin Eye Res.* 2018;66:17-48.

13. Cervino NG, Elias-Costa AJ, Pereyra MO, Faivovich J. A closer look at pupil diversity and evolution in frogs and toads. *Proc R Soc B.* 2021;288:20211402. https://doi.org/10.1098/rspb.2021.1402.

14. Whitaker, BR. The amphibian eye. In: Wright KM, Whitaker BR, eds. *Amphibian Medicine and Captive Husbandry.* Malabar, FL: Kreiger Publishing; 2001:245-252.

15. Williams DL. Ocular surface biology and disease in amphibians. *Vet Clin North Am Exot Anim Pract.* 2019;22:97-107. https://doi.org/10.1016/j.cvex.2018.08.008.

16. Franz-Odendaal TA. Skeletons of the eye: an evolutionary and developmental perspective. *Anat Rec.* 2020;303(1):100-109.

17. Fritsches K, Warrant E. Vision. In: Wyneken J, Lohmann K, Musick J, eds. *Biology of Sea Turtles.* Vol. III. Boca Raton, FL: CRC Press; 2003.

18. Granda AM, Dvorak CA. 1977. Vision in turtles. In: Crescitelli F, ed. *The Visual System in Vertebrates.* Berlin, Germany: Springer Verlag; 1977;451-495.

19. Schwab IR. From eye spots to eye shine. *Br J Ophthalmol.* 2000;84:1214-1216.

20. Collin SP. Behavioural ecology and retinal cell topography. In: Archer SN, Djamgoz MBA, Loew E, Partridge JC, Vallerga S, eds. *Adaptive Mechanisms in the Ecology of Vision.* Amsterdam, The Netherlands: Kluwer Academic Publishers; 1999:509-535.

21. Caprette CL, Lee MSY, Shine R, et al. The origin of snakes (Serpentes) as seen through eye anatomy. *Biol J Linn Soc Lond.* 2004;81(4):469-482.

22. Ott M. Visual accommodation in vertebrates: mechanisms, physiological response and stimuli. *J Comp Physiol A* 2006;192:97. https://doi.org/10.1007/s00359-005-0049-6.

23. Yovanovich CAM, Koskela SM, Nevala N, et al. The dual rod system of amphibians supports colour discrimination at the absolute visual threshold. *Phil Trans R Soc.* 2017;B372: 20160066.http://dx.doi.org/10.1098/rstb.2016.0066.

24. Hart NS, Coimbra JP, Collin SP, Westhoff G. Photoreceptor types, visual pigments, and topographic specializations in the retinas of hydrophiid sea snakes. *J Comp Neurol.* 2012;520(6):1246-1261.

25. Davies WL, Cowing JA, Bowmaker JK, et al. Shedding light on serpent sight: the visual pigments of henophidian snakes. *J Neurosci.* 2009;29: 7519-7525.

26. Nagloo N, Collin SP, Hemmi JM, Hart NS. Spatial resolving power and spectral sensitivity of the saltwater crocodile, *Crocodylus porosus*, and the freshwater crocodile, *Crocodylus johnstoni*. *J Exp Biol.* 2016;219(9):1394-1404.

27. Hauzman E, Bonci DM, Ventura DF. Retinal topographic maps: a glimpse into the animals' visual world. In: Heibockel T, ed. *Sensory Nervous System. IntechOpen.* 2018;1:101-26.

28. Martinez PS, Plummer CE. Ophthalmology of Crocodilia: alligators, crocodiles, caimans, and gharials. In: Montiani-Ferreira F, Moore BA, Ben-Shlomo G, eds. *Wild and Exotic Animal Ophthalmology.* New York, NY: Springer; 2022. https://doi.org/10.1007/978-3-030-71302-7.

29. Oliver, LJ, Salmon M, Wyneken J, et al. Retinal anatomy of hatchling sea turtles: anatomical specializations and behavioral correlates. *Mar Freshw Behav Physiol.* 2000;33:233-248.

30. el Hassni M, M'Hamed SB, Repérant J, Bennis M. Quantitative and topographical study of retinal ganglion cells in the chameleon (*Chameleo chameleon*). *Brain Res Bull.* 1997;44(5):621-625.

31. Wilhelm M, Straznicky C. The topographic organization of the retinal ganglion cell layer of the lizard *Ctenophorus nuchalis*. *Arch Histol Cytol.* 1992;55(3):251-259.

32. Collin SP. A web-based archive for topographic maps of retinal cell distribution in vertebrates. *Clin Exp Optom.* 2008;91(1):85-95.

33. Inger R, Bennie J, Davies TW, Gaston KJ. Potential biological and ecological effects of flickering artificial light. *PLoS ONE.* 2014;9(5):e98631. https://doi.org/10.1371/journal.pone.0098631.

34. Young M, Salmon M, Forward R. Visual wavelength discrimination by the loggerhead turtle, *Caretta caretta*. *Biol Bull.* 2012;222:46-55.

35. Granda, AM, Haden KW. Retinal oil globule counts and distributions in two species of turtles: *Pseudemys scripta elegans* (Wied) and *Chelonia mydas mydas* (Linnaeus). *Vision Res.* 1970;10:79-84.

36. Vorobyev M. Coloured oil droplets enhance colour discrimination. *Proc Biol Sci.* 2003;270:1255-1261.

37. Brischoux F, Pizzatto L, Shine R. Insights into the adaptive significance of vertical pupil shape in snakes. *J Evol Biol.* 2010;23(9):1878-1885.

38. Sivak JG, Levy B, Weber AP, Glover RF. Environmental influence on shape of the crystalline lens: the amphibian example. *Exp Biol.* 1985;44(1):29-40.

39. Porter WR, Witmer LM. Vascular patterns in iguanas and other squamates: blood vessels and sites of thermal exchange. *PLoS ONE.* 2015;10(10):e0139215.

40. Meyer-Rochow VB, Wohlfahrt S, Ahnelt PK. Photoreceptor cell types in the retina of the tuatara (*Sphenodon punctatus*) have cone characteristics. *Micron.* 2005;36:423-428.

41. Wake MH. *Hyman's Comparative Vertebrate Anatomy.* Chicago, IL: University of Chicago Press; 1979.

42. Wyneken J. *The Anatomy of Sea Turtles.* U.S. Department of Commerce NOAA Technical Memorandum NMFS-SEFSC-470; 2001.

43. Reese AM. *The Alligator and Its Allies.* New York, NY: G. P. Putnam and Sons; 1915.

44. Coke RL, Couillard NK. Ocular biology and diseases of old world chameleons. *Vet Clin North Am Exot Anim Pract.* 2002;5(2):275-285.

45. Wake, MH The comparative morphology and evolution of the eyes of caecilians (Amphibia, Gymnophiona). *Zoomorphology.* 1985;105:277-295. https://doi.org/10.1007/BF00312059.

Section 32

Anesthesia, Surgery, and Imaging

Sue Horton, DVM;
Stacey Wilkinson, DVM, DABVP (Reptile and Amphibian Practice)
Session Chairs

Rhinotomy, Rhinoscopy and Treatment of an Oronasocutaneous Fistula in a Sulcata Tortoise (*Centrochelys sulcata*)

Mariana Sosa-Higareda, MVZ,
Hugues Beaufrère, Dr med vet, PhD, DABVP (Avian Practice), DipECZM (Avian), Dipl ACZM,
Kathryn Phillips, DVM, Dipl ACVR,
David Sanchez-Migallon Guzman, LV, MS, DipECZM (Avian), DipECZM (Small Mammal), Dipl ACZM

Session #2663

Affiliation: From the William T. Pritchard Veterinary Medical Teaching Hospital (Sosa-Higareda, Phillips), Department of Medicine and Epidemiology (Guzman, Beaufrère) School of Veterinary Medicine University of California-Davis, Davis, CA 95616, USA.

Note: Abstract format selected by author(s). See Front Matter for description of conference proceedings manuscripts submitted as scientific abstracts.

Abstract: A 10-year-old female sulcata tortoise (*Centrochelys sulcata*) was presented for evaluation of a non-healing wound on her left rostrum. On physical examination there was a 5 × 5 mm defect dorsal to the left nares, with compacted plant material inside. Intraoral examination revealed a 3 × 3 mm keratin defect at the level of the premaxilla with visible compacted plant material. A computed tomographic study was performed and revealed an oronasocutaenous fistula with compression nasomaxillary fractures and left-sided rhinitis and food impaction. PCR from a nasal flush sample for intranuclear coccidia was negative. The patient underwent general anesthesia for fistula repair. A rhinotomy to extend the rostral defect was performed to allow plant material removal, lavage, and nasal cavity exploration. A rhinoscopy through the rhinotomy site revealed hyperemic and inflamed mucosa. Flushing was performed through the oral defect. The rhinotomy site was covered with a hydrocolloid bandage. The fistula was plugged with a dental bis-acryl composite resin, which was also applied over the intraoral keratin defect and along the maxillary tomium. Antibiotics and analgesics were administered post-operatively. The patient recovered well from the procedure. On evaluation 8 weeks after surgery, the resin covering the rhinotheca's keratin defect at the occlusal surface remained in place and the rostral defect had significantly contracted with adequate healing. The tortoise will eventually outgrow the resin covering the rhinothecal defect; however, it is expected that repeated procedures will be able to provide long-term management. There are limited reports of oronasocutaneous fistulas in chelonians[1,2] and in this case, trauma is the inciting cause.

References

1. Boyer TH. Differential diagnoses by clinical signs: chelonians. In: Divers, SJ, Stahl SJ, eds. *Mader's Reptile and Amphibian Medicine and Surgery.* 3rd Ed. St. Louis, MO: Elsevier Saunders; 2019: 1266-1275.

2. Innis CJ, Garner MM, Johnson AJ, et al. Antemortem diagnosis and characterization of nasal intranuclear coccidiosis in Sulawesi tortoises (*Indotestudo forsteni*). *J Vet Diagn.* 2007;19(6):660-667.

Novel Shell Fracture Stabilization Technique in a Blanding's Turtle

Nicole T. Mueller, DVM, and
Kenneth J. Waliszewski, DDS, MS (Prosthodontics)

Session #2709

Affiliation: From BluePearl Specialty and Emergency Pet Hospital, 2100 W Silver Spring Dr, Glendale, WI 53209, USA (Mueller), Department of Graduate Prosthodontics, School of Dentistry, Marquette University, 1801 W Wisconsin Ave, Milwaukee, WI 53201, USA (Waliszewski).

Note: Abstract format selected by author(s). See Front Matter for description of conference proceedings manuscripts submitted as scientific abstracts.

Abstract: An adult Blanding's turtle (*Emydoidea blandingii*) presented with multiple carapacial fractures and one horizonal plastron fracture. After multiple attempts at plastron fracture stabilization, a novel repair resulted from collaboration with a prosthodontist and neuroscientist. Bone anchor screws commonly used to anchor cannulae to the skull in research rodents were placed on either side of the fracture. A pliable sheet of light-curable composite resin was molded over the fracture and onto the imbedded screw heads. The resin was exposed to high-intensity light, resulting in a rigid splint. This material allowed the patient to return to its semi-aquatic captive habitat the following day and resume normal behaviors. Complete healing was confirmed after 2.5 years. The Wisconsin Department of Natural Resources applied a radio transmitter and the patient was released. This approach is less invasive than traditional bone screw and plate fixation, reducing the risk of pathogenic contamination or injury to underlying bone and intracoelomic structures. After resin and screws are removed, there is no delay to release for epithelization of screw holes. The smooth, moldable resin can be used anywhere on the shell, minimizes drag, and eliminates the risk of material adhering to fixation devices. Disadvantages of this technique center around the lack of fracture line access to evaluate healing or provide wound care. In theory, the resin does not allow for natural shedding of scutes; however, in this patient there was no evidence of subsequent dermatitis or osteomyelitis. This case highlights the value of creative collaboration for successful outcomes.

Comparison of Anesthetic Protocols Using Isoflurane versus Total Injectable Anesthesia Using Hydromorphone, Alfaxalone, and Midazolam for Radiotelemetry Surgery in Northern Non-Coastal Pinesnakes (*Pituophis melanoleucus melanoleucus*)

Gregory B. Walth, DVM,
Jessica R. Comolli, DVM,
Bryan D. Hudson,
H. Nicole Trenholme, DVM, MS, DACVECC, DACVAA,
Stephen J. Divers, BVetMed, DZooMed, DipECZM (Herpetology),
DipECZM (Zoo Health Management), Dipl ACZM, FRCVS

Session #2603

Affiliation: From Small Animal Medicine and Surgery Department (Zoological Medicine), University of Georgia College of Veterinary Medicine, Athens, GA 30602, USA (Walth, Comolli, Divers), Clemson University Department of Forestry and Environmental Conservation, Clemson, SC 29634, USA (Hudson), Department of Large Animal Medicine, University of Georgia College of Veterinary Medicine, Athens, GA 30602, USA (Trenholme).

Note: Abstract format selected by author(s). See Front Matter for description of conference proceedings manuscripts submitted as scientific abstracts.

Abstract: Volatile anesthetic gases utilized to maintain a surgical plane of anesthesia are often associated with significant cardiopulmonary (CP) effects in reptiles. Total injectable anesthesia may be a safe alternative with less CP sequelae and shorter recovery. This retrospective study compared two anesthetic protocols for surgical implantation or removal of a radiotransmitter in non-coastal northern pinesnakes (*Pituophis melanoleucus melanoleucus*). Subjects were maintained with isoflurane alone or in combination with other anesthetics (ISO) or total injectable anesthesia using hydromorphone 0.5-1 mg/kg, alfaxalone 10-20 mg/kg, and midazolam 1-2 mg/kg (HAM). Ten snakes and 16 anesthetic episodes were included in the study, with eight anesthetic episodes included in each treatment group. Six snakes underwent both ISO and HAM treatments, with a minimum of 39 weeks between treatments. All patients were monitored with Doppler for heart rate (HR) and capnography for end tidal carbon dioxide ($ETCO_2$). All data were evaluated for normality with a Shapiro-Wilk test and analyzed with a two-sample one tailed t test with alpha set at 0.05. Both HR and $ETCO_2$ were significantly lower in ISO compared to HAM ($P < 0.05$). Apnea was not noted in the majority (6/8) of HAM. The total anesthetic time through recovery was significantly longer ($P < 0.05$) in ISO (168.9 ± 32.7 min) compared to HAM (113.3 ± 27.4 min) with no statistical difference in total surgery times ($P = 0.058$). Total injectable anesthesia appears to be safe and effective for healthy northern pinesnakes undergoing brief and routine surgical procedures.

Use of a 45° Oblique Radiograph in Tortoises with Femoral Head Ostectomy

Ian Kanda, RVT, VTS (Exotics), and
Thomas Boyer, DVM, DABVP (Reptile and Amphibian Practice)

Session #2715

Affiliation: From Pet Hospital of Penasquitos, 9888 Carmel Mountain Rd, Ste F, San Diego, CA 92129, USA.

Note: Abstract format selected by author(s). See Front Matter for description of conference proceedings manuscripts submitted as scientific abstracts.

Abstract: Two injured wild Californian desert tortoises (*Gopherus agassizii*) were presented by their monitoring field biologist with hind leg lameness. One tortoise also had contaminated and infected bite wounds of the left hip. Traditional dorsoventral, lateral, and cranio-caudal radiographs produced poor detail of the affected coxofemoral joint due to overlapping of soft tissue and the bones of the shell, pelvis and femur. Based on a view described in birds,[1] a 45° oblique craniodorsal-caudoventral view offered a vastly superior view of the coxofemoral joint. This view was used to prescribe a femoral head ostectomy and verify successful completion following surgery.

References

1. Visser M, Hespel A-M, de Swarte M, Bellah JR. Use of a caudoventral-craniodorsal oblique radiographic view made at 45° to the frontal plane to evaluate the pectoral girdle in raptors. *J Am Vet Med Assoc.* 2015;247(9): 1037-1041.

Novel Use of Oblique Radiographic Views to Assess the Pectoral and Pelvic Girdles of Turtles

Miranda Torkelson, DVM, CWR, CertAqV

Session #2607

Affiliation: From the Wildlife Rehabilitation Center of Minnesota, 2530 Dale St N, Roseville, MN 55113, USA.

Note: Abstract format selected by author(s). See Front Matter for description of conference proceedings manuscripts submitted as scientific abstracts.

Abstract: Turtles often present to wildlife rehabilitation centers for trauma secondary to motor vehicle collisions, with reported percent of total turtle admissions ranging from 42-63%.[1-3] While external injuries to the shell are straightforward to diagnose and various techniques for repair are well described,[4] determining the extent of internal injuries is more difficult. Injuries to the pectoral and pelvic girdles in turtles are challenging to assess on standard radiographic views due to the superimposition of bones. Due to the fusion of their spine and shell, all motion in walking and swimming is reliant upon movements of limbs and mobility within the shoulders and pelvis.[5,6] In addition to their effect on motility pelvic fractures may inhibit mating and nest building as well as decrease the size of the pelvic inlet, which may cause dystocia. The novel use of caudoventral-craniodorsal and cranioventral-caudodorsal 45° oblique radiographic views in aquatic pond turtles at the Wildlife Rehabilitation Center of Minnesota has improved differentiation of bones in the pelvic and pectoral girdle, improving our diagnostic ability. Using standard radiographic views, 50% (58/117) of painted turtles (*Chrysemys picta)* with caudal carapace fractures had suspected injuries to the pelvis, however the specific injury was difficult to determine. After starting to use oblique views, 100% (24/24) of painted turtles with caudal carapace fractures were found to have pelvic involvement, and these injuries were able to be specifically diagnosed. Improved visualization has enabled us to make informed decisions regarding prognosis and treatment options for turtle patients.

References

1. Sack A, Butler E, Cowen P, Lewbart GA. Morbidity and mortality of wild turtles at a North Carolina wildlife clinic: a 10-year retrospective. *J Zoo Wildl Med.* 2017;48:716-724. https://doi.org/10.1638/2016-0053.1

2. Schrader GM, Allender MC, Odoi A. Diagnosis, treatment, and outcome of Eastern box turtles (*Terrapene carolina carolina*) presented to a wildlife clinic in Tennessee, USA, 1995-2007. *J Wildl Dis.* 2010;46:1079-1085. https://doi.org/10.7589/0090-3558-46.4.1079

3. Rivas AE, Allender MC, Mitchell M, Whittington JK. Morbidity and mortality in reptiles presented to a wildlife care facility in central Illinois. *Human-Wildl Interact.* 2014;8:78-87. https://doi.org/10.26077/34wh-vh39

4. Norton TM, Fleming GJ, Meyer J. Shell surgery and repair. In: *Mader's Reptile and Amphibian Medicine and Surgery.* 3rd Ed. St. Louis, MO: Elsevier Saunders; 2019:1116-1126.

5. Rivera G, Rivera ARV, Dougherty EE, Blob RW. Aquatic turning performance of painted turtles (*Chrysemys picta*) and functional consequences of a rigid body design. *J Exp Biol.* 2006;209:4203-4213. https://doi.org/10.1242/jeb.02488

6. Pace CM, Blob RW, Westneat MW. Comparative kinematics of the forelimb during swimming in red-eared slider (*Trachemys scripta*) and spiny softshell (*Apalone spinifera*) turtles. *J Exp Biol.* 2001;204:3261-3271. https://doi.org/10.1242/jeb.204.19.3261

Diagnostic Accuracy of Oblique Radiographic Views for Experimentally Induced Fractures of Pectoral and Pelvic Girdles in Red-Eared Slider Cadavers (*Trachemys scripta elegans*)

Grace H. Chung, BVetMed, MRCVS,
Carrie Kuzma, DVM, Dipl ACVR,
Nicola Di Girolamo, DMV, MSc (EBHC), PhD, DipECZM (Herpetology), Dipl ACZM,
João Brandão, LMV, MS, DipECZM (Avian)

Session #2704

Affiliation: From Department of Veterinary Clinical Sciences, College of Veterinary Medicine, Oklahoma State University, Stillwater, OK 74078, USA.

Note: Abstract format selected by author(s). See Front Matter for description of conference proceedings manuscripts submitted as scientific abstracts.

Abstract: Chelonian pelvic and pectoral girdles are difficult to assess on standard radiographic views due to superimposition. This study evaluated the diagnostic accuracy of two novel oblique radiographic views (craniodorsal-caudoventral oblique made at 45°to the frontal plane (Cr45D-CdV) and caudodorsal-cranioventral oblique made at 45° to the frontal plane (Cd45D-CrV)) to diagnose pelvic and pectoral girdle fractures. Eighteen red-eared slider (*Trachemys scripta elegans*) cadavers had seven radiographic views (DV, right lateral, cranio-caudal, Cr45D-CdV, Cd45D-CrV, the obliques taken on blocks) taken before and after injury to a random combination of the left and/or right pelvic and/or pectoral girdles. A blinded reviewer assessed each randomly numbered radiograph. The sensitivity of the Cr45D-CdV view with and without the block was 50% and the sensitivity of the Cd45D-CrV was 46.4% without a block and 64.3% with the block. The sensitivity of the standard views was relatively poor at 21.4% for the DV view, 0% for the lateral view, and 10.7% for the cranio-caudal view. The specificity of the Cr45D-CdV, Cd45D-CrV, right lateral, and cranio-caudal was 100%, while the specificity of the DV was 94.7% and the specificity of the Cr45D-CdV and Cd45D-CrV views with the block were 98.9% and 99.5% respectively. The addition of these novel oblique radiographic views to the standard set of radiographs can help diagnose pelvic and pectoral girdle injuries of chelonians.

Diagnostic Radiology of Aquatic Freshwater Chelonians

Miranda Torkelson, DVM, CWR, CertAqV

Session #2608

Affiliation: From Wildlife Rehabilitation Center of Minnesota, 2530 Dale St N, Roseville, MN 55113, USA.

Abstract: Radiographic assessment of chelonians can be more difficult than for other taxonomic groups due to their bony shell as well as superimposition of structures. To further complicate matters, there is conflicting information in the literature on radiographic anatomy of chelonians, and this anatomy differs between species. In this master class, standard radiographic views for turtles (dorsal-ventral, cranial-caudal, and lateral) are reviewed as well as the novel use of caudoventral-craniodorsal and cranioventral-caudodorsal oblique radiographic views to assess the pectoral and pelvic girdles. Assessment of respiratory, gastrointestinal, reproductive, and skeletal structures are discussed. Concepts are illustrated using case-based examples from common snapping turtles and painted turtles in rehabilitation.

Introduction

Aquatic freshwater chelonians are commonly presented to wildlife rehabilitation centers, with admission causes including trauma (e.g., collisions, animal attack, other), fish hooks, and infections (respiratory infections, aural abscess).[1,2] External injuries are relatively straightforward to diagnose on exam, however internal trauma and underlying illnesses are more difficult to assess. Radiographs are a valuable diagnostic tool for chelonians, however due to their bony shell techniques and interpretation are different than with other species. A thorough knowledge of chelonian anatomy is vital to the interpretation of diagnostic imaging and can be obtained through the study of written resources.[3] There is conflicting information in the literature likely stemming from species differences, so clinicians should seek to solidify their anatomy knowledge through necropsy exams of different species.

Techniques

Standard radiographic views

Standard radiographic views for turtles include dorsoventral, lateral, and cranial-caudal views. These views allow adequate imaging of lung fields, axial skeleton, and appendages.[4] Dorsoventral views are taken with a vertical beam, whereas lateral and cranial-caudal views are best taken with a horizontal beam. If a horizontal beam is not available, the turtle can be positioned with radiolucent objects to take the same views with a vertical beam, however resulting images should be interpreted with caution as shifting of organs or any coelomic fluid can alter the appearance of structures. The patient should always be placed as close as possible to radiographic plate to minimize distortion. Appropriate positioning is vital to visualize structures and assess asymmetry. Often, it is possible to take these views without sedation. Dorsoventral views can be taken with limbs in either a flexed or extended position, the combination of which can aid in the assessment of limbs and joints. In order to take lateral views, it is helpful to elevate patient by placing plastron on top of a radiolucent object to facilitate extension of head and limbs.[5]

Oblique radiographic views

It can be difficult to assess the pectoral and pelvic girdle in turtles due to superimposition of bones. Oblique radiographic views are commonly used in humans and adapted to avian species in order to better assess the shoulder girdle.[6,7] We have adapted this avian H view (caudoventral-craniodorsal 45 degree oblique) to turtles as well as the perpendicular oblique angle, which we have termed the A view (cranioventral-caudodorsal 45 degree oblique). These oblique views separate the different bones of the shoulder and pelvis in turtles and have resulted in improved diagnostic imaging and interpretation.

"Plop" radiographs

In wildlife rehabilitation, it is often important to get initial diagnostics at admission to make triage and treatment decisions, however patients may not be stable enough for a complete diagnostic workup. We utilize "plop" radiographs at admission for all chelonian patients; named because the patient is plopped down on the plate and a dorsoventral whole body radiograph is taken. Plop rads are very useful for ruling out metallic foreign bodies such as fishhooks, checking for eggs, surveying for major fractures, and looking for abnormal gas patterns in GI tract. It is important to not overinterpret these radiographs; positioning is typically not symmetric and they do not include orthogonal views.

Sedation and anesthesia

Depending on the radiographic views necessary, any additional procedures, and patient compliance, sedation or anesthesia may need to be considered. Protocols for sedation and anesthesia vary by species.[8,9]

Application and Interpretation

Skeletal

Radiographs can be used to assess for injury or damage to the bones of the skull and jaw. Proper positioning is imperative with skull radiographs. Survey radiographs of the entire turtle can be helpful as an initial assessment of the skull, however if there is concern about this area collimated orthogonal skull radiographs and potentially oblique views should be taken. Sedation or anesthesia may be required. Skull radiographs are indicated when there is concern for a jaw or facial fracture, head trauma, neurological signs, or an aural abscess (to rule out secondary bone lysis).

The spine is difficult to fully assess in turtle radiographs due to the fusion of the spine with the shell, however the spine should be evaluated on dorsoventral and lateral views for any step lesions, abnormal spaces, or fractures to the shell that may cross the spine as these may indicate severe spinal trauma.

It can be a challenge to assess the pectoral and pelvic girdles in turtles due to superimposition of bones. Oblique views can be used to separate these different bones for improved clarity. The long bones of the limbs should be assessed with orthogonal views and joints assessed in flexed and extended positions, similar to other species. Radiographs can also be useful to assess bone quality in chelonians. The digits and distal bones of the extremities are often the first to show poor density and mineralization on radiographs in cases of malnutrition.

Respiratory

Radiographs are an essential diagnostic and monitoring tool for chelonians with respiratory disease. Three views are necessary to fully visualize the lung fields: right and left lateral and cranial-caudal. Turtles can have unilateral (more common) or bilateral lung pathology.[10] Pneumonia is characterized by pulmonary consolidation and increased opacity on radiographs. Positioning is important because a clear view of the lung fields can be obscured by head and limbs tucked into shell. After initial diagnosis, typically the author repeats radiographs monthly to monitor for improvement vs progression. It is important to use the same positioning and technique with recheck radiographs to be able to compare opacities.

Gastrointestinal

Typically, coelomic detail is poor due to superimposition of the shell on internal organs; however, dorsoventral radiographs can rule out GI abnormalities such as foreign bodies and ileus. The most common foreign bodies found in wild aquatic chelonians are fishhooks, which can be diagnosed and localized or ruled out with radiographs.

A large amount of gas within the intestines can indicate decreased GI motility or ileus. Barium studies can be utilized in chelonian species. The author uses a dose of 5 ml/kg of a barium solution tubed into the stomach. A dorsoventral radiograph should be taken immediately after barium administration and every 30 minutes until barium enters intestines as GI transit tends to slow. Radiographs can then be checked every 1-2 hours the rest of the day then 1-2 times/day until barium exits the body. Complete clearance of barium can take days to weeks in chelonians with decreased GI motility.

Radiographs can be used to confirm placement of esophageal tubes for nutrition or medication administration. Some feeding tubes are more radio-opaque than others; if necessary, barium can be placed in the tube to provide more contrast. It is important to confirm appropriate placement of tube in radiographs with head and neck both extended, as the tube may appear to be far enough in the GI tract on a flexed view but still only be in the esophagus.

Reproductive

Radiographic monitoring is vital for gravid female chelonians to ensure eggs develop and are laid properly. Monthly dorsoventral radiographs are recommended for gravid chelonians in care to check how many eggs remain and monitor for any abnormalities. Eggs should be free from deformities, be able to pass through the pelvic inlet, and move towards the pelvic inlet over time. If a chelonian is not laying appropriately, the environment should be assessed, and a full diagnostic workup is recommended.

References

1. Rivas AE, Allender MC, Mitchell M, Whittington JK. Morbidity and mortality in reptiles presented to a wildlife care facility in central Illinois. *Human-Wildl Interact*. 2014;8(1):78-87. doi:10.26077/34wh-vh39.

2. Sack A, Butler E, Cowen P, Lewbart GA. Morbidity and mortality of wild turtles at a North Carolina wildlife clinic: a 10-year retrospective. *J Zoo Wildl Med*. 2017;48(3):716-724. doi:10.1638/2016-0053.1.

3. O'Malley B. Tortoises and turtles. In: *Clinical Anatomy and Physiology of Exotic Species*. St. Louis, MO: Elsevier Saunders; 2005:41-56. doi:10.22233/9781905319909.14.

4. Pees M. Reptiles 3.1 radiographic investigation. In: *Diagnostic Imaging of Exotic Pets*. Germany: Schlütersche; 2011:310-333.

5. Holmes SP, Divers SJ. Radiography: chelonians. In: *Mader's Reptile and Amphibian Medicine and Surgery*. 3rd Ed. St. Louis, MO: Elsevier Saunders; 2019:514-527.

6. Kornguth PJ, Salazar AM. The apical oblique view of the shoulder: its usefulness in acute trauma. *Am J Roentgenol*. 1987;149(1):113-116. doi:10.2214/ajr.149.1.113.

7. Visser M, Hespel AM, de Swarte M, Bellah JR. Use of a caudoventral-craniodorsal oblique radiographic view made at 45° to the frontal plane to evaluate the pectoral girdle in raptors. *J Am Vet Med Assoc*. 2015;247(9):1037-1041. doi:10.2460/javma.247.9.1037.

8. Schnellbacher RW, Shepard M. Sedation. In: *Mader's Reptile and Amphibian Medicine and Surgery*. 3rd Ed. St. Louis, MO: Elsevier Saunders; 2019:441-446.

9. Mans C, Sladky KK, Schumacher J. General anesthesia. In: *Mader's Reptile and Amphibian Medicine and Surgery*. 3rd Ed. St. Louis, MO: Elsevier Saunders; 2019:447-464.

10. Farrow CS. Turtles. In: *Veterinary Diagnostic Imaging: Birds, Exotic Pets and Wildlife*. St. Louis, MO: Elsevier Mosby; 2009:403-413.

Validation of Coelomic Veterinary Point-of-Care Ultrasound in Multiple Snake Species: Pilot Study

Kaleigh Eichel, BKI, DVM,
Søren Boysen, DVM, DACVECC,
Serge Chalhoub, BSc, DVM, DACVIM (Small Animal Internal Medicine),
Douglas P. Whiteside, DVM, DVSc, Dipl ACZM, DipECZM (Zoo Health Management)

Session #2721

Affiliation: From the Department of Veterinary Clinical and Diagnostic Sciences (Boysen, Chalhoub) and the Department of Ecosystem and Public Health (Whiteside), Faculty of Veterinary Medicine (Eichel), University of Calgary, 11877 85th Street NW, Calgary, AB, Canada T3R 1J3, Wilder Institute-Calgary Zoo, 1300 Zoo Road NE, Calgary, AB, Canada T2E 7V6 (Whiteside).

Note: Abstract format selected by author(s). See Front Matter for description of conference proceedingsmanuscripts submitted as scientific abstracts.

Abstract: Background: Veterinary point-of-care ultrasound (VPOCUS) has been underutilized in zoological medicine with no published protocol for use in snakes. Methods: This study assessed the utility of a standardized protocol for coelomic VPOCUS in 13 healthy snakes of both sexes: 5 colubrid and 4 boid species. With snakes in dorsoventral position, sonographic exams followed this protocol to locate and evaluate coelomic organs and answer binary questions about organ ratios and presence of intracoelomic fluid. Results: Intracoelomic fluid was not present in any snake. Lung parenchyma and expansion with inhalation was appreciated in all snakes. Air sac movement was appreciated although the caudal border of the caudal air sac was not identified in 2/13 snakes and was misidentified as significantly more cranial by 8% ± 9% (mean ± 1SD) of snake length compared to radiographic location ($P = 0.018$). There was no significant difference between colubrid and boid organ ratios. Mean right atrium to pulmonary trunk ratio was 1.27:1 ± 0.47:1. Mean ventricle wall to lumen ratio was 1.11:1 ± 0.2. Mean hepatic vein diameter to liver parenchyma height was 19.8% ± 7.4%. Mean gallbladder diameter to snake diameter was 35% ± 13%. Mean vena cava diameter to aorta diameter ratio was 0.97:1 ± 0.17:1. Conclusion: There is further need to validate normal organ ratios and determine the significance of these ratios in diseased snakes. This standardized VPOCUS protocol can be used to systematically evaluate coelomic organs for potential indicators of critical pathologies and presence of intracoelomic fluid in everyday practice.

Development and Standardization of Cardiac Diagnostic Parameters of the Bearded Dragon (*Pogona vitticeps*)

Jessica Conner, BS,
Bart Huber, DVM,
Curtis Eng, DVM

Session #2626

Affiliation: From the Department of Exotic, Zoo and Wildlife Medicine, College of Veterinary Medicine, Western University of Health Sciences Pomona, CA 91766, USA (Conner, Eng), Animal Medical Center of Corona, 91 E Grand Blvd STE 102, Corona, CA 92879, USA (Huber).

Note: Abstract format selected by author(s). See Front Matter for description of conference proceedings manuscripts submitted as scientific abstracts.

Abstract: Heart disease in reptiles has not been well described, clinical signs are fairly non-specific and could very well be diagnostically under-represented.[1-7] As reptiles live longer it is probable that the prevalence of heart disease will continue to rise. One of the many diagnostic tools in detecting cardiovascular disease is measurement of heart size. While the vertebral heart score is well known and quite established in cats and dogs, there are no such standards for diagnostic cardiac evaluation in reptiles. This retrospective study aims to define normal heart parameters in a small population of bearded dragons. Using past cases that required radiology and computed tomography, the long axis of the heart was measured at its longest section; the short axis was then measured at the tallest section in a perpendicular segment from the long axis. The measurements of the sum these heart axes were compared to vertebral sections, lizard length, pectoral and pelvic girdle lengths. For further comparison, the same measurements were recorded from bearded dragons previously diagnosed with cardiac pathologies. The sum of the long and short axes demonstrated poor correlation when compared to girdle lengths; however, the same measurements showed moderate correlation when compared to lizard length ($r = 0.59$), showing promise for future comparison to vertebral length. Overall, bearded dragons with diagnosed clinical disease had an average heart size 21% larger than those without cardiac pathology. Descriptive data analysis provided through this study will begin the process of creating a set of standard heart parameters for future clinical cardiac diagnoses.

References

1. Banzato T, Hellebuyck T, Van Caelenberg A, et al. A review of diagnostic imaging of snakes and lizards. *Vet Rec.* 2013;173(2):43-49.

2. Banzato T, Selleri P, Veladiano IA, Zotti A. Comparative evaluation of the cadaveric and computed tomographic features of the coelomic cavity in the green iguana (*Iguana iguana*), black and white tegu (*Tupinambis merianae*) and bearded dragon (*Pogona vitticeps*). *Anat Histol Embryol.* 2013;42(6):453-460.

3. Bucy DS, Sanchez-Migallon Guzman D, Zwingenberger AL. Ultrasonographic anatomy of bearded dragons (*Pogona vitticeps*). *J Am Vet Med Assoc.* 2015;246(8):868-876.

4. Simone-Freilicher E, Sullivan P, Quinn R, et al. Two cases of congestive heart failure in lizards. *Proc ExoticsCon.* 2015;505-509.

5. Mitchell MA. Reptile cardiology. *Vet Clin North Am Exot Anim Pract.* 2009;12(1):65-79.

6. Silverman S. Diagnostic imaging. In: *Mader's Reptile and Amphibian Medicine and Surgery.* 2nd Ed. St. Louis, MO: Elsevier Saunders; 2006;471-502.

7. Silverman S, Sanchez-Migallon Guzman D, Stern J, et al. Standardization of the two-dimensional trans-coelomic echocardiographic examination in the central bearded dragon (*Pogona vitticeps*). *J Vet Cardiol.* 2016;18(2):168-178.

Section 33

Clinical Pathology

Shane Simpson, BVSc(Hons), GCM(VP)
Session Chair

Performance of a Veterinary Portable Glucose Meter for Blood Glucose Concentration Measurement in Red-Eared Sliders (*Trachemys scripta elegans*)

Nicola Di Girolamo, DMV, MS, PhD, DipECZM (Herpetology), Dipl ACZM,
Michael Bonas,
Marina Liles, DVM

Session #2723

Affiliation: From the Department of Clinical Sciences, College of Veterinary Medicine, Oklahoma State University, Stillwater, OK 74078, USA.

Note: Abstract format selected by author(s). See Front Matter for description of conference proceedings manuscripts submitted as scientific abstracts.

Abstract: There is limited knowledge on the accuracy and precision of portable blood glucose meters (PBGMs) for measurement of blood glucose in reptiles. We aimed to evaluate a veterinary PBGM for measuring blood glucose concentration in red-eared sliders (*Trachemys scripta elegans*) and to evaluate the effect of sample characteristics on its performance. Blood samples were obtained from 20 red-eared sliders from the jugular vein. The blood was placed in a lithium-heparin tube. Two strips of a veterinary PBGM (AlphaTrak 2, Zoetis) on a dog setting were run in rapid succession. Immediately afterwards, the blood was spun, and plasma separated. Plasma was frozen and shipped for glucose measurement via a hexokinase analyzer. Passing-Bablok regression and Bland-Altman analysis were built to determine agreement between the results obtained with the PBGM and the reference test. The veterinary PBGM drastically overestimated blood glucose concentration, with a mean overestimation of 28.0 mg/dl (95% CI: 16.7 to 45.0, Limits of Agreement: 5.8 to 50.3 mg/dl). The disagreement was proportional, increasing at higher blood glucose concentrations. The mean difference was 23.1 mg/dl in the samples with PCV between 10 and 19% (n = 9), 30.8 mg/dl in the samples with PCV between 20 and 29% (n = 7), and 34.4 mg/dl in the samples with PCV between equal or over 30% (n = 4). Intra-assay coefficients of variations, a measure of precision, were on average 4.5%, ranging from 0 to 17.4%. In conclusion, the use of the veterinary PBGM evaluated in this study in red-eared sliders may alter both treatment and diagnostic decisions because of the overestimation of glucose concentrations.

Determination of Blood Biochemistry Reference Intervals in Eastern Collared Lizards (*Crotaphytus collaris*)

Brandon J. Louth, BS, DVM,
J. Jill Heatley, DVM, MS, DABVP (Avian Practice), DABVP (Reptile and Amphibian Practice) Dipl ACZM,
Karen E. Russell, PhD, DVM, DACVP (Clinical Pathology)

Session #2693

Affiliation: From Research Boulevard Pet and Bird Hospital, Austin, TX 78759, USA (Louth), College of Veterinary Medicine & Biomedical Sciences, Texas A&M University, College Station, TX 77843, USA (Heatley, Russel).

Note: Abstract format selected by author(s). See Front Matter for description of conference proceedingsmanuscripts submitted as scientific abstracts.

Abstract: Eastern collared lizards (*Crotaphytus collaris*) are growing in popularity in the exotic pet trade and breeding industry in the United States and around the world; however, clinical biochemistry reference data to support interpretation of health and disease for these species remains lacking.[1-20] This study evaluated ninety-two apparently healthy collared lizards (*Crotaphytus* spp.), including eighty-seven eastern collared lizards (*Crotaphytus collaris*) and five Dickerson's collared lizards (*Crotaphytus dickersonae*). Blood samples were collected from the ventral coccygeal tail vein and tested with the Abaxis VetScan VS2 biochemical analyzer. While lizard weight and SVL (snout to vent length) were normally distributed, most biochemical analytes, except albumin and total protein were not. Many analytes were affected, albeit slightly, by the following variables assessed: age, SVL, body condition score, gravidity, sex, oviposition, apparently healthy, color/locality; however, with the exception of calcium, phosphorous, and albumin for gravid or immediately post oviposition females, other values remained within the generated reference interval. Limitations in this study were the limited range of bile acid capability of the VetScan VSA ABAXIS Analyzer unable to successfully read the majority of the values below a BA of 35 μmol/L and analyzer limitations for efficient evaluation of albumin and globulin. Methods used in previous studies on BA may be needed to properly evaluate BA levels in the *Crotaphytus* genus. When compared to biochemistry values of the green iguana, plumed basilisk, Allen Cays rock iguana, and Ricord's iguana in the same infraorder Iguania from the western hemisphere, collared lizards shared similar ranges in some analytes, but several differences were noted that can be significant when used for determination of clinical diseases or abnormalities.

References

1. Burt CE. The synonymy, variation, and distribution of the collared lizard, *Crotaphytus collaris*. *Occas Papers Mus Zool.* 1928;196:1-19.

2. Templeton AR, Robertson RJ, Brisson J, Strasburg J. Disrupting evolutionary processes: the effects of habitat fragmentation on collared lizards in the Missouri Ozarks. *Proc Natl Acad Sci.* 2001;98(10):5426-5432.

3. Ballinger RE, Hipp TG. Hematological variations in the collared lizard, *Crotaphytus collaris* (Sauria:Iguanidae). *Copeia.* 1985;3:782-784.

4. Plasman M, Reynoso VH, Nicolás L, Torres R. Multiple colour traits signal performance and immune response in the Dickerson's collared lizard *Crotaphytus dickersonae. Behav Ecol Sociobiol.* 2015;69: 765-775.

5. Louth BJ, Heatley JJ. Husbandry and captive propagation of eastern collared lizards (*Crotaphytus collaris*). *J Herpetol Med Surg.* 2020;30(4):1-14.

6. Cooper W, Ferguson G. Estrogenic priming of color change induced by progesterone in the collared lizard, *Crotaphytus collaris. Herpetologica.* 1973;19(2): 107-111.

7. Redrobe S, MacDonald J. Sample collection and clinical pathology of reptiles. *Vet Clin North Am Exot Anim Pract.* 1999;2(3):709-730.

8. Friedrichs KR, Harr KE, Freeman KP, et al. ASVCP reference interval guidelines: determination of de novo reference intervals in veterinary species and other related topics. *Vet Clin Pathol.* 2012;41(4):441-53.

9. Fitch HS. A comparative study of loss and regeneration of lizard tails. *J Herpetol.* 2003;37(2):395-399.

10. Campbell TW. Clinical pathology. In: Mader DR, ed. *Reptile Medicine and Surgery.* Philadelphia, PA: W.B. Saunders; 1996;248-257.

11. Baird TA, Hews DK. Hormone levels in territorial and non-territorial male collared lizards. *Physiol Behav.* 2007;92(4):755-763.

12. Divers SJ, Redmayne G, Aves EK. Haematological and biochemical values of 10 green iguanas (*Iguana iguana*). *Vet Rec.* 1996;138(9):203-205.

13. Harr KE, Alleman AR, Dennis PM, et al. Morphologic and cytochemical characteristics of blood cells and hematologic and plasma biochemical reference ranges in green iguanas. *J Am Vet Med Assoc.* 2001;218(6): 915-921.

14. Dallwig RK, Paul-Murphy J, Thomas C, et al. Hematology and clinical chemistry values of free-ranging basilisk lizards (*Basiliscus plumifrons*) in Costa Rica. *J Zoo Wildl Med.* 2011;42(2):205-213.

15. James SB, Iverson V, Greco V, Raphael B. Health assessment of allen cays rock iguana, *Cyclura cychlura inornata. J Herpetol Med Surg.* 2006;16(3):93-97.

16. Maria R, Ramer J, Reichard T, et al. Biochemical reference intervals and intestinal microflora of free-ranging Ricord's iguanas (*Cyclura ricordii*). *J Zoo Wildl Med.* 2007;38(3):414-419.

17. McCain SL, Flatland B, Schumacher JP, et al. Comparison of chemistry analytes between 2 portable, commercially available analyzers and a conventional laboratory analyzer in reptiles. *Vet Clin Pathol.* 2010;39(4):474-479.

18. Hagey LR, Vidal N, Hofmann AF, Krasowski MD. Evolutionary diversity of bile salts in reptiles and mammals, including analysis of ancient human and extinct giant ground sloth coprolites. *BMC Evol Biol.* 2010;10:133

19. Greenacre CB, Flatland B, Souza MJ, Fry MM. Comparison of avian biochemical test results with abaxis vetscan and hitachi 911 analyzers. *J Avian Med Surg.* 2008;22(4):291-299.

20. McGuire JA. Phylogenetic systematics of crotaphytid lizards (Reptilia: Iguania: Crotaphytidae). *Bull Carnegie Mus Nat Hist.* 1996;32:1-143.

Determining A Plasma Retinol Reference Range in Leopard Geckos (*Eublepharis macularius*)

Brandon J. Louth, BS, DVM,
J. Harrison Wellford, DVM,
J. Jill Heatley, DVM, MS, DABVP (Avian Practice), DABVP (Reptile and Amphibian Practice), Dipl ACZM

Session #2697

Affiliation: From Research Boulevard Pet and Bird Hospital, Austin, TX 78759, USA (Louth, Wellford), College of Veterinary Medicine & Biomedical Sciences, Texas A&M University, College Station, TX 77843-4474, USA (Heatley).

Note: Abstract format selected by author(s). See Front Matter for description of conference proceedings manuscripts submitted as scientific abstracts.

Abstract: The leopard gecko (*Eublepharis macularius*) is a species of gecko native to the Middle East and is a popular pet in the exotic community around the world. There is minimal knowledge known on ranges of retinol in this or closely related species.[1-3] As this species becomes more popular in the exotic community more medical conditions are being identified that can be assisted in diagnosis with the establishment of ranges of the different essential analytes. This study evaluated and characterized an interval for plasma retinol activities of the leopard gecko. The geckos in this study were maintained according to published husbandry species standards. Blood of 40 clinically healthy geckos was drawn from the ventral abdominal vein using a 25-gauge needle on a 1-cc syringe. Blood was spun down using a centrifuge to obtain plasma samples, then placed in a retinol standard and analyzed using LCMS analysis for retinol concentrations. With the selected sample size there was a statistically significant range analyzed for the plasma retinol values sampled to help determine a reliable reference interval for leopard geckos and similar species to allow clinicians the ability to interpret biochemistry concentrations from collected blood samples. The novel procedure used in this study can be used to evaluate plasma retinol values with a sample volume as low as 50 μl.

References

1. Cojean O, Alberton S, Froment R, et al. Determination of leopard gecko (*Eublepharis macularius*) packed cell volume and plasma biochemistry reference intervals and reference values. *J Herpetol Med Surg.* 2020;30(3):156-164.

2. Ferguson GW. Indoor husbandry of the panther chameleon *Chamaeleo furcifer pardalis*: effects of dietary vitamins A and D and ultraviolet irradiation on pathology and life-history traits. *J Zoo Biol.* 1996;15(3): 279-299.

3. Frutchey KP. Plasma levels of vitamins A and E in marine turtles (*Chelonia mydas* and *Caretta caretta*). 2004. STARS. Retrospective Theses and Dissertations. University of Central Florida. 4580.

Evaluating Blood Lactate and Glucose Obtained Through Point-of-Care Testing in Reptiles

Bronwyn Koterwas, BVM&S MRCVS FHEA

Session #2653

Affiliation: From Dick Vet Rabbit and Exotics Practice, Hospital for Small Animals, Royal (Dick) School of Veterinary Studies, University of Edinburgh, Easter Bush Campus, Edinburgh, Scotland, EH25 9RG, United Kingdom.

Abstract: Point-of-care (POC) hand-held blood lactate and glucose monitors are used in small animal medicine to help determine severity of disease, metabolic derangement and response to treatment. These parameters are integral in the determination of systemic inflammatory response syndrome (SIRS) as well as the Rule of 20, key aspects of a critically ill patient that should be evaluated daily. By quickly assessing response to treatment, clinicians are able to either adjust treatment or prevent suffering through euthanasia.

Reptile blood samples cannot be run on common in-house hematologic and biochemical table top analysers. There can be a delay of several days when awaiting blood results from external laboratories leading to either postponed treatment or instigation of stop-gap treatments. Quicker in-house testing through the use of POC machines could help clinicians adjust clinical plans in the moment and choose more appropriate treatment modalities. Lactose dehydrogenase (LDH), the enzyme responsible for conversion of pyruvate to lactate, is a fairly non-specific marker of illness but has been shown to elevate with cardiac and renal disease in reptiles. LDH is tested routinely over lactate because it minimally degrades over time. It is reasonable to conclude that lactate, when tested at time of sample collection, would also be elevated in these cases as is seen in mammals. Similarly, blood glucose levels may increase in cases of pancreatic, liver and gastrointestinal disease as well as neoplastic disorders. Blood glucose and lactate are simple, minimally invasive tests that could have a large impact on course of treatment and therefore overall prognosis.

Introduction

Point-of-care (POC) hand-held blood lactate and glucose monitors are commonly used in small animal medicine to help determine severity of disease, metabolic derangement and response to treatment. Despite anatomic and physiologic differences, particularly with metabolism, the authors hypothesized that blood glucose and lactate would elevate in a similar fashion in debilitated reptiles.

Normal reference ranges of specific biochemical analytes exist for many reptile species,[1] but these can vary with season, reproductive status, age and size resulting in inconsistent correlation with clinical disease.[2] Monitoring less specific parameters, such as glucose and lactate, may allow clinicians to have a better idea of general, systemic disease state in their patients. Establishing reliable trends in a rapid fashion would therefore help select more appropriate treatment modalities and alleviate suffering in reptiles.

To our knowledge, there has been no study investigating lactate and blood glucose levels using point of care monitors in companion reptiles in the clinical setting despite many publications suggesting further research is indicated. If the value of these tests can be established, they would provide a quick, minimally invasive and inexpensive methodology for collecting real evidence in the effective treatment of reptiles.

Study Design

Ethical approval

Ethical approval for this study was granted through the University of Edinburgh's Veterinary Ethical Review Committee (VERC 114.20). Blood samples are regularly collected to aid in diagnosis of illness as well as response to treatment. This is particularly valuable in species, such as reptiles, which can harbour quite advanced yet subclinical disease. In this study, excess blood which may have been discarded (such as from the hub of a syringe or butterfly needle tubing) was used to test for blood glucose and lactate using an in-house POC machine.

Clinical sampling during this procedure was only done in the best interest of the individual based on actual/real time clinical need and performed in a minimally invasive nature. Repeat sampling was not performed for the sole purpose of obtaining further samples for research or future testing. Rather than being disposed of, residual material (as may be collected during routine blood sampling) was utilized.

The owners of these reptiles signed a standard consent form already in use by the Hospital for Small Animals (HfSA). Owner consent for samples taken as part of a diagnostic evaluation at the HfSA was documented on the standard HfSA consent form which allows excess samples to be retained for future research when a clinically relevant sample is collected.

Sample technique

Samples were taken from the ventral coccygeal vein, jugular vein, subcarapacial sinus, and less commonly intra-cardiac in sedated snakes, per standard accepted practices. Recommended values for total blood sample volumes of 0.4-0.8 ml/100 g of body weight with 0.4-0.5 ml/100 g in debilitated animals were used.[3]

Other clinical parameters such as physical examination results, body weight, body condition, age, reproductive status, body temperature also were documented when known. These were analyzed in conjunction with other blood results, diagnostic tests and treatment outcomes in order to evaluate the overall usefulness of point of care testing in the clinical setting.

Equipment

An AlphaTrak2 veterinary blood glucose monitor (Zoetis, UK), placed on the canine setting was used for all blood glucose values. 0.3 μl of whole blood was required for this blood lactate was obtained using a StatStrip Xpress lactate meter (Nova Biomedical; USA). Sampling commenced immediately after blood collection and samples were either taken directly from a plain syringe or a vial containing lithium heparin aniticogulant. The Statstrip Xpress lactate meter gave results within 13 seconds and required a minimum blood volume of 0.6 μl with a range of 0.3-20 mmol/L.

Results

Blood samples were obtained from a variety of reptiles; most commonly lizards and tortoises. Blood lactate values ranged from 0.8-9.0 mmol/L while blood glucose varied between 3.1 and 30.4 mmol/L.

Discussion

POC blood glucose and lactate monitors rely on very small volumes (often 0.2-0.7 μl) of whole blood to deliver an immediate result; their use has been validated in small animal medicine.[4,5] These parameters also are integral in the determination of systemic inflammatory response syndrome (SIRS) and as well as the Rule of 20,[6] key aspects of a critically ill patient that should be evaluated daily. By quickly assessing response to treatment, clinicians are able to either adjust treatment or prevent suffering through euthanasia.[5,7]

Reptile blood sample values have not been validated on common in-house hematologic and biochemical tabletop analyzers. Most clinicians must send samples to external laboratories that have the appropriate equipment and expertise in processing these samples and interpreting results. There can be a delay of several days when awaiting blood results leading to either postponed treatment or instigation of stop-gap treatments. Quicker, in-house testing through the use of validated POC machines would help clinicians choose more appropriate treatment modalities.

Lactose dehydrogenase (LDH), the enzyme responsible for conversion of pyruvate to lactate, is a fairly non-specific marker of illness but has been shown to be elevated with cardiac and renal disease in reptiles.[8] LDH is tested routinely over lactate because it minimally degrades over time. It is reasonable to conclude that lactate, when tested at time of sample collection, would also be elevated in these cases as is seen in mammals.

Similarly, blood glucose levels may increase in cases of pancreatic, liver and gastrointestinal disease as well as neoplastic disorders.[9] Clinicians may rely on elevated blood glucose in rabbits as an aid in determining whether a gastrointestinal obstruction can be medically or surgically managed.[10] Blood glucose and lactate are simple, minimally invasive tests that can have a large impact on course of treatment and therefore overall prognosis. Finding a correlation between changes to blood glucose and lactate along with pathology would likewise assist clinicians in caring for their reptile patients as well.

Conclusions

Normal reference ranges of specific biochemical analytes exist for many reptile species, but these can vary with season, reproductive status, age and size resulting in inconsistent correlation with clinical disease. Monitoring less specific parameters, such as glucose and lactate, may allow clinicians to have a better idea of general, systemic disease state in their patients. Establishing reliable trends in a rapid fashion could help select more appropriate treatment modalities and alleviate suffering in reptiles. As this study accumulates larger numbers of patient data, it is hoped that values of blood glucose and lactate can be validated on POC monitors.

References

1. Klaphake E, Gibbons P, Sladky K, Carpenter JW. Reptiles. In: Carpenter JW, ed. *Exotic Animal Formulary.* 5th Ed. Elsevier; 2018:81-166.

2. Eatwell K, Hedley J. Reptile haematology and biochemistry. *InPractice.* 2014;36(1):34-42.

3. Divers SJ. Diagnostic techniques and sample collection. In: Divers SJ, Stahl SJ, eds. *Mader's Reptile and Amphibian Medicine and Surgery.* 3rd Ed. St. Louis, MO: Elsevier; 2019:405-408.

4. Thorneloe C, Bedard C, Boysen S. Evaluation of a hand held lactate analyser in dogs. *Can Vet J.* 2007;48(3):283-288.

5. Di Mauro, FM, Schoeffler, GL. Point of care measurement of lactate. *Top Companion Anim Med.* 2016;31(1):1-44.

6. Kirby R. An introduction to SIRS and the rule of 20. In: Kirby R, Linklater A, eds. *Monitoring and Intervention in the Critically Ill Small Animal: The Rule of 20.* Ames, IA: Wiley-Blackwell; 2017:1-9.

7. Kohen C, Hopper K. Lactate monitoring. In: Drobatz KJ, Hopper K, Rozanski E, Silverstein DC, eds. *Textbook of Small Animal Emergency Medicine.* Ames, IA: Wiley-Blackwell; 2018:1005-1010.

8. Schilliger L, Girling S. Cardiology. In: Divers SJ, Stahl SJ, eds. *Mader's Reptile and Amphibian Medicine and Surgery.* 3rd Ed. St. Louis, MO: Elsevier; 2019;683.

9. Heatley JJ, Russell KE. Clinical chemistry. In: Divers SJ, Stahl SJ, eds. *Mader's Reptile and Amphibian Medicine and Surgery.* 3rd Ed. St. Louis, MO: Elsevier; 2019:319-330.

10. Harcourt-Brown FM, Harcourt-Brown SF. Clinical value of blood glucose measurement in pet rabbits. *Vet Rec.* 2012;170(26):674.

Leukocyte Counts in Red-Sided Garter Snakes (*Thamnophis sirtalis parietalis*) During Spring Emergence in Drought

Kristen K2 Larson, BS, BA,
Brianna Beechler, DVM, PhD, BA, PhD,
Robert T. Mason, PhD, BA,
Christopher R. Friesen, PhD, BS

Session #2707

Affiliation: From Carlson College of Veterinary Medicine, Oregon State University, Magruder Hall 700 30th St. Corvallis, OR 97330, USA (Larson, Beechler), Department of Zoology, Oregon State University, Cordley Hall 3029, Corvallis, OR 97330, USA (Mason), School of Earth, Atmospheric and Life Sciences, University of Wollongong NSW 2522 Australia (Friesen).

Abstract: During spring in Manitoba, Canada, 120 red-sided garter snakes (*Thamnophis sirtalis parietalis*) were collected (Oregon State University ACUP no. 4317) and 200-300 μl blood was taken from caudal tail veins for leukocyte analyses. Sodium heparin was used as an anticoagulant (less than 5 μl/syringe). Manitoba experienced a drought during the study; all snakes were collected from den sites that had no water within 500 meters. Two treatment groups included: snakes provided humidity (43) a desiccant (42), and a control group with no manipulation of humidity (35). Reference ranges for WBC counts were determined based on animals lacking heterophil and lymphocyte reactivity; total WBC count range resulted in 3400-22115 WBC/μl. Newly emerged males (seasonal virgins, NeM) did not show a significant difference in total WBCs compared to males previously mated within the season (unpaired T-test, P value = 0.5274, t = 0.6353), suggesting that mated status does not alter overall leukocyte counts in males. NeM did show a higher mean heterophil count at 15.2 cells/μl over 10.29 cells/μl (P = 0.02, t = 2.381). Treatments with humidity levels did not affect WBC counts whether individuals were kept in humid or dry conditions (one-way ANOVA, F = 1.030, P = 0.3603). Females had slightly higher WBC counts with a mean of 11,231 WBC/μl compared to males with a mean of 9,026 WBC/μl (P-value = 0.02823, t = 2.21) (Fig 1). Females also had increased heterophils (17.5 cells/μl, males: 11.7 cells/μl, P = 0.001, t = 3.24), and monocytes (63.3 cells/μl, males: 53.7, P = 0.01, t = 2.56) but decreased lymphocyte counts compared to males (14.8 cells/μl, males: 29.7 cells/μl, P = 0.0001, t = 3.99).

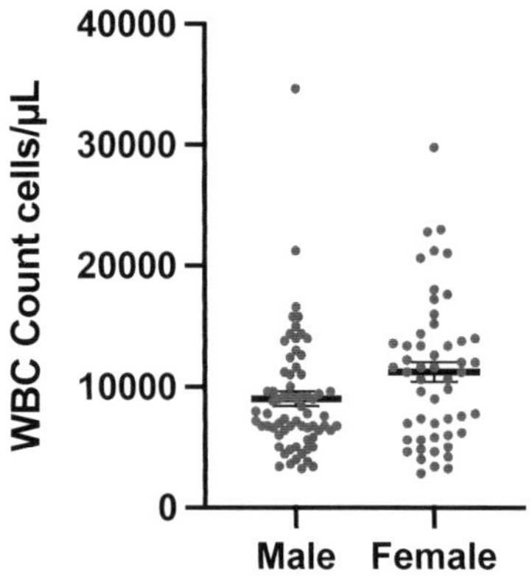

Figure 1. White blood cell counts of male and female red-sided garter snakes.

Association of Plasma Metabolites and Diagnostic Imaging with Hepatic Lipidosis in Bearded Dragons (*Pogona vitticeps*)

Trinita K. Barboza, DVM, DVSc,
Leonardo Susta, DVM, PhD, DACVP,
Alex zur Linden, DVM, Dipl ACVR,
Sara Gardhouse, DVM, Dipl ACZM,
Hugues Beaufrère, DVM, PhD, Dipl ACZM, DABVP (Avian Practice), DipECZM (Avian)

Session #2545

Affiliation: From Department of Clinical Sciences, Cummings School of Veterinary Medicine, Tufts University, North Grafton, MA 01536, USA (Barboza), Department of Pathobiology (Susta) and Clinical Studies (zur Linden), Ontario Veterinary College, University of Guelph, Guelph, Ontario, Canada, Department of Clinical Sciences, College of Veterinary Medicine, Kansas State University, Manhattan, KS 66502, USA (Gardhouse), Department of Veterinary Medicine and Epidemiology, UC Davis School of Veterinary Medicine, Davis, CA 95616, USA (Beaufrère).

Note: Abstract format selected by author(s). See Front Matter for description of conference proceedings manuscripts submitted as scientific abstracts.

Abstract: The objective of this study was to evaluate the association between plasma metabolites, biochemical analytes, diagnostic imaging findings, and hepatic lipidosis in bearded dragons. Fourteen bearded dragons (*Pogona vitticeps*) with varying severity of hepatic lipid accumulation (with and without hepatic lipidosis) were included. Animals underwent coelomic ultrasound, computed tomography (CT) scans, and coelioscopic hepatic biopsies. Clinical pathology tests included lipidologic tests, hepatic biomarkers, and mass spectrometry-based metabolomics. Hounsfield units on CT were negatively associated with increased hepatic vacuolation, while ultrasound and gross evaluation of the liver were not reliable. On biochemistry, low beta-hydroxybutyric-acid (BHBA) was significantly associated with hepatic lipidosis. Metabolomics and lipidomics data found BHBA and succinic acid to be potential biomarkers for diagnosing moderate-to-severe hepatic lipidosis. These findings provide information on the antemortem assessment of hepatic lipidosis in bearded dragons and paves the way for further research in diagnosis and treatment of this disease.

Clinical and Pathologic Findings of Urolithiasis in Green Iguanas (*Iguana iguana*): A Review of 12 Cases (2011-2021)

Anthony J. Cerreta, DVM,
Krista A. Keller, DVM, Dipl ACZM,
Sara M. Gardhouse, DVM, DABVP (Exotic Companion Mammal Practice), Dipl ACZM,
David Sanchez-Migallon Guzman, LV, MS, DipECZM (Avian), DipECZM (Small Mammal), Dipl ACZM

Session #2719

Affiliation: From Department of Veterinary Clinical Medicine, College of Veterinary Medicine, University of Illinois at Champaign-Urbana, Urbana, IL 61801, USA (Carreta, Keller), Department of Clinical Sciences, College of Veterinary Medicine, Kansas State University, Manhattan, KS 66506, USA (Gardhouse), Department of Medicine and Epidemiology, School of Veterinary Medicine, University of California-Davis, Davis, CA 95616, USA (Guzman).

Note: Abstract format selected by author(s). See Front Matter for description of conference proceedings manuscripts submitted as scientific abstracts.

Abstract: Urinary calculi have been previously reported in squamates, although the current literature is limited to single case reports.[1,3-5] The aims of this retrospective study were to evaluate the clinical and pathologic findings in green iguanas (*Iguana iguana*) with urolithiasis. The medical records of green iguanas presented between 1996 and 2021 were reviewed and twelve cases (seven females, five males) with confirmed urolithiasis were identified from three institutions. Cases were included after surgical confirmation (7/12), postmortem examination (3/12), or advanced imaging (2/12). Average (min-max range) age at time of presentation was 12-22) years. Clinical signs included lethargy (8/12), tenesmus (6/12), and hyporexia (5/12). A moderate leukocytosis consistent with inflammation was found in three cases.[2] Moderate hyperphosphatemia and hypercalcemia, suggestive of renal disease was seen in two cases.[2] Radiographs were performed in ten animals. A single circular mineralized lamellar structure was noted in the caudal coelom in eight cases, and multiple smaller ovoid mineralized stones in two cases. Seven animals had urolith removal via cystotomy (6/7) or cystectomy (1/6). Average (min-max range) survival time after surgery was 41.3 (2-96) months. Three cases had calculi analyzed and were composed of 100% salt(s) of uric acid. Aerobic bacterial culture from bladder wall collected during surgery was performed in two cases, resulting in pure growth of *Klebsiella* spp. (1/2) and *Salmonella* spp., *Pseudomonas* spp., and *Enterococcus* spp. (1/2). Urate urolith prevention strategies, including diet and husbandry risk factors should be evaluated further in green iguanas under human care.

References

1. Birke LL, Cespedes AM, Schachner ER, Lailvaux SP. Cystic calculus in a laboratory-housed green anole (*Anolis carolinensis*). *Comp Med.* 2017;67(2):112-115.

2. Harr KE, Alleman AR, Dennis PM, et al. Morphologic and cytochemical characteristics of blood cells and hematologic and plasma biochemical reference ranges in green iguanas. *J Am Vet Med Assoc.* 2001;218(6): 915-921.

3. Kwantes LJ. Surgical correction of cystic urolithiasis in an iguana. *Can Vet J.* 1992;33(11):752-753.

4. Lightfoot T. Bladder necrosis secondary to cystic calculus in a green iguana. *Exotic DVM.* 1999;1:29-33.

5. Wolf KN, Troan BV, DeVoe R. Chronic urolithiasis and subsequent cystectomy in a San Esteban Island chuckwalla, *Sauromalus varius*. *J Herpetol Med Surg.* 2008;18(3):106-112.

Blood Culture and Synovial Fluid Analysis of Rescued Sea Turtles in Taiwan (2019-2021)

Hsiang-Pei Su, DVM,
Chau-Hwa Chi, DVM, PhD,
Pin-Huan Yu, DVM, PhD

Session #2636

Affiliation: From Institute of Veterinary Clinical Sciences, School of Veterinary Medicine, National Taiwan University, No.1, Sec. 4, Roosevelt Road, Taipei, 10617, Taiwan, R.O.C.

Note: Abstract format selected by author(s). See Front Matter for description of conference proceedings manuscripts submitted as scientific abstracts.

Abstract: Septic arthritis is one of several frustrating illnesses in sea turtle rehabilitation due to its unclear pathogenesis, delayed-onset during rehabilitation, long-term treatment requirement and poor prognosis.[1-4] Radiographs, blood culture and arthrocentesis have been used as diagnostic tools for suspected cases; however, previous studies showed that bacteremia could exist in clinically heathy reptiles. To the best of our knowledge, prevalence of bacteremia and the characteristics of normal synovial fluid in sea turtles has not yet been established. During 2019-2021, we enrolled 16 rescued sea turtles in the study under inclusion criteria: normal limb motor functions, no radiographic evidence of joint lesion and no hematologic or plasma biochemistry evidence of systemic infection/illness. Bacterial culture of the blood and synovial fluid from shoulder joints as well as qualitative analysis of the synovial fluid were obtained from these sea turtles. The results revealed bacterial culture-negative of both blood and synovial fluid in arthritis-free sea turtles. Cytology of the normal synovial fluid were dominated by mononuclear cells and occasionally a few heterophils or red blood cells. This study provides information on bacterial culture of blood and normal synovial fluid characteristics in rescued sea turtles in Taiwan. The results of the study might improve the diagnosis of septic arthritis of sea turtles in the future.

References

1. Cruciani B, Shneider F, Ciccione S, et al. Management of polyarthritis affecting sea turtles at Kélonia, the Reunion Island Sea Turtle Observatory (2013-17). *J Wildl Dis.* 2019;55(2):455-461.

2. Innis CJ, Braverman H, Cavin JM, et al. Diagnosis and management of *Enterococcus* spp infections during rehabilitation of cold-stunned Kemp's Ridley turtles (*Lepidochelys kempii*): 50 cases (2006–2012). *J Am Vet Med Assoc.* 2014;245(3):315-323.

3. Powell AL, Tuxbury KA, Cavin JM, et al. Osteomyelitis in cold-stunned Kemp's Ridley sea turtles (*Lepidochelys kempii*) hospitalized for rehabilitation: 25 cases (2008-2018). *J Am Vet Med Assoc.* 2021;259(10): 1206-1216.

4. Waugh L, Ramachandran A, Talent S, et al. Survey of aerobic and anaerobic blood cultures in free-ranging Western ratsnakes (*Pantherophis obsoletus*). *J Herpetol Med Surg.* 2017;27(1-2):44-47.

Section 34

Potpourri

Stacey Wilkinson, DVM, DABVP (Reptile and Amphibian Practice)
Session Chair

Turtle Euthanasia Demystified

Renée Schott, DVM, CWR

Session #2522

Affiliation: From The Wildlife Rehabilitation Center of Minnesota, 2530 Dale St. N, Roseville, MN 55113, USA.

Abstract: Many veterinarians struggle with turtle euthanasia. Many have heard of euthanized turtles moving after being declared dead, which has led to odd turtle euthanasia practices in an attempt to prevent reanimation. There is abundant literature on turtle physiology and consciousness, yet it has not been applied it to euthanasia practices. This highly referenced and interesting master class reveals why turtles are so hard to euthanize and provides a protocol that will minimize turtle suffering during the euthanasia process and ensure non-reversible death.

Previously published in Schott R. Turtle euthanasia demystified. Proc ExoticsCon. 2021;483-489.

Introduction

Reptiles have all the anatomy necessary to feel pain and suffer and therefore should be treated with the same humane standards as other species, which includes providing them with proper analgesics.

The term euthanasia is derived from the Greek terms eu meaning well or good, and Thanatos meaning death. It is defined as ending the life of an animal in a way that eliminates or minimizes pain and distress.[1,2] The technique employed should result in rapid loss of consciousness followed by cardiac or respiratory arrest and, ultimately, a loss of brain function. Actually determining the pain/distress of a method of euthanasia can be very difficult because as humans we will never fully know/understand the subjective experience of the animal.[3] We use our best judgement; paddling, vocalizations, convulsions before apparent loss of consciousness is obviously suffering. We also use our knowledge of physiology and assume suffering in the absence of behaviors if a physiologic process theoretically leads to suffering. Pain, distress and suffering are subjective affective experiences that are perceived in the brain. Therefore, one must be conscious and alive to suffer. General rule: a gentle death that takes longer is preferable to a rapid but more distressing death and conversely, if all other methods are equally humane, the quickest method should be chosen.[4] Taking a life is abhorrent to many, we often forget this and need to keep it in mind.

Turtle Physiology Pertinent to Euthanasia Methods

Natural history

Turtles are poikilothermic (survive at a wide temperature range), and ectotherms (get their body heat from the outside world). Painted and snapping turtles have been seen brumating simply in the water (not in mud) but also in the mud, muskrat burrows/lodges, solo, together, seeps/springs, cattail stands, under banks, under logs, out in open water, etc, so many different environments.[5]

Anoxia tolerant brains

In commonly studied reptile species, anoxia tolerance appears to vary widely to opposite ends of the spectrum, although it is possible that if more species were investigated anoxia tolerance would be found as a relatively continuous spectrum.[5] Varying degrees of anoxia tolerance has been documented in chelonian species. These differences can be attributed in part to several metabolic factors, including the degree of metabolic depression, the effectiveness of lactic acid buffering, and the capacity of glycogen stores.[6] Electrical and biochemical brain activity is believed to persist in the anoxic chelonian brain for a considerable period of time.[7] Some species of freshwater turtles (*Trachemys* and *Chrysemys* genera) can survive many weeks without oxygen (commonly used as model animals for vertebrate anoxia tolerance).[8] Some freshwater turtle brains can survive anoxia for months (ie under the water during winter),[9] *Trachemys scripta* which can withstand complete anoxia for days at room temperature to weeks in winter hibernation.[5,6] Even at room temperature, 24 hours of anoxia and re-oxygenation results in no evident loss of neurons.[10]

During anoxic periods, these turtles may appear comatose due to brain and body down regulation: Turtles can down-regulate their brains to become virtually comatose.[9] One mechanism is to extend anoxic survival is entrance into a state of deep reversible hypo-metabolism.[11] Energy demand is reduced to meet the energy supplied by anaerobic glycolysis.[11] Energy demanding processes are greatly suppressed.[6,11] Ion channels, and neurotransmitters are downregulated by multiple intertwined mechanisms, suppression of action potentials, protein regulation, reviewed here.[11]

Neuroprotection (and why they don't die of reperfusion): A variety of protective mechanisms are activated at the molecular level in anoxic turtle brain, many of which may not only protect against damage under anoxic conditions, but also ameliorate oxidative stress when oxygen is restored.[11] These include increases in heat shock proteins, anti-apoptotic factors, the MAP kinases, antioxidants and modulation of the p53 pathway. Unlike the mammalian brain, which shows an overproduction of reactive oxygen species (ROS) following hypoxia or ischemia/reperfusion, the turtle brain appears to suppress ROS production upon re-oxygenation.[12]

Latitudes matter: In the northern parts of their ranges in North America, turtles may spend more than half of their lives in an overwintering state.[5] Ontario turtles tolerate anoxia much better vs Nebraska turtles.[13] The three northern subspecies are very anoxia-tolerant, and can remain responsive for more than 150 days in anoxic water, but the southern subspecies can survive only about half that time, having a more rapid fall in pH.[5] Northern turtles are better able to buffer accumulated lactate vs southern turtles, and thus survive longer in anoxic water.[14] Aquatic turtles with far northern ranges are more anoxia tolerant than southern animals even within a single species (probably because of the need to survive potentially long periods under ice or in hypoxic mud).[5]

Lactic acid buffering

Lactic acid can be both buffered and sequestered by bone. Chelonians may have particularly good capacity for buffering since there is so much well vascularized bone in the shell.[5,6]

Brain cell regeneration

It has been shown that after significant ischemic damage, within 3 weeks *T. scripta* has evidence of neuronal reproduction.[15] There are robust repair mechanisms present including neurogenesis of damaged brain cells.[12]

Ages in some species

Baby painted turtles hatch in fall, stay in ground, and live off yolk and come out in the spring.[13] Interestingly, hatchlings of many turtle species survive in the nest in their first winter through freeze tolerance or super-cooling;[16] however, in general hatchlings are far less able to tolerate anoxic submergence than adults. This could be due to their incomplete shell development, as the shell is important for lactate buffering.[5]a major portion of their life cycle has received relatively little attention until recently - namely what turtles are doing, and how they are doing it, during the winter. In the northern parts of their ranges in North America, turtles may spend more than half of their lives in an overwintering state. In this review, I emphasise the ecological aspects of overwintering among turtles, and consider how overwintering stresses affect the physiology, behaviour, distributions, and life histories of various species. Sea turtles are the only group of turtles that migrate extensively, and can therefore avoid northern winters. Nevertheless, each year a number of turtles, largely juveniles, are killed when trapped by cold fronts before they move to safer waters. Evidently this risk is an acceptable trade-off for the benefits to a population of inhabiting northern developmental habitats during the summer. Terrestrial turtles pass the winter underground, either in burrows that they excavate or that are preformed. These refugia must provide protection against desiccation and lethal freezing levels. Some burrows are extensive (tortoise genus Gopherus Snapping turtles emerge the same year they were laid as they cannot overwinter in the nest.

Unacceptable Euthanasia Methods

Hypothermia

Placing an animal in a freezer to cause death or dropping it into liquid nitrogen is unacceptable. These methods as standalone are not acceptable for euthanasia of reptiles, as ice crystals formed in the body are presumed to be painful, and the turtle might not die (consider physiology previously described).

Isoflurane

Because isoflurane[2] (and other gas anesthetics such as halothane or sevoflurane) have a noxious smell and can irritate respiratory epithelium, animals may hold their breath. In particular, turtles (because of the physiology mentioned above) can hold their breath for hours or days when exposed to isofurane.[17] For this reason, isoflurane as a euthanasia method is not recommended.[18]

Hyperthermia

Raising the temperature above the critical temperature of the species. This method is not acceptable for euthanasia of turtles.

Suffocation and drowning

Depriving animals of oxygen. This method is not acceptable for euthanasia of turtles.

Traditional Veterinary Euthanasia

Anesthesia

Anesthesia should first be induced in any turtle being euthanized. Whatever doses of whatever drugs you use for surgery in the species you are seeing, double it. For painted turtles and snapping turtles, this author uses alfaxalone (20 mg/kg) IM or SC and then 15 minutes later follows that up with 10-20 mg/kg propofol IV. Turtles are often too small to hit a vein awake so anesthetize with alfaxalone and then consider the occipital venous sinus.

Sodium pentobarbital: Sodium pentobarbital is a barbiturate developed in 1930s as an anesthetic[19] that has fallen out of use as an anesthetic. Currently, it is overdosed to cause death/euthanasia, and has been recommended as an effective and human method of euthanasia in reptiles.[3]

Sodium pentobarbital comes in two general formulations (with and without phenytoin), both of which are ~390 mg/ml and pHs are between 9.6-13. Routes of administration include:

- Intravenous
  - Ideally, pentobarbital is given intravenous (100 mg/kg).
  - May be difficult and likely stressful to try to hit a vein in an awake animal as most reptile pheblotomy is done blindly.
  - But the pH may cause venous endothelium cell damage on contact (painful).
- Intracardiac injection may only be used on a fully anaesthetized animal as this is very painful.[18]
- Intra-lung is presumed to be painful and should not be performed.
- Intracoelomic administration is described; however, all formulations of sodium pentobarbital are caustic, and some advise they not be given extravascularly as it is perceived to be painful.[20] In addition, it is slower acting.[18]
- Per os: administering sodium pentobarbital, typically via gavage tubing as it tastes bitter, has been used by some veterinarians but may be prohibited in certain states. Barbiturates have high oral bioavailability and can reach the brain quite quickly. A disadvantage is this method will be slower than IV so in a moribund animal may be ineffective.
- Intramuscular/subcutaneous should not be used as they are not effective and may cause pain.[18]

Sodium pentobarbital works in stages in mammals. These stages are:

Stage 1. Voluntary excitement

- Bloodstream → heart → cerebral cortex
- Begins to lose consciousness/coordination, hyperesthesia, loses voluntary motor, loss of superficial pain → disorientation, possible movement

Stage 2. Involuntary excitement

- Bloodstream → heart → cerebral cortex → cerebrum
- Loss of consciousness → possible uncontrolled motor activity (paddling, vocalizations)

Stage 3. Surgical anesthesia

- Bloodstream → heart → cerebral cortex → cerebrum → cerebellum (~5 sec)
- Cannot feel pain, doesn't respond to visual/auditory, reflexes begin to disappear
- Because of depressed metabolic rates in sick, dying, cold or otherwise depressed turtles, many individual turtles may not circulate the pentobarbital quickly enough to advance from this stage to Stage 4. If this happens, the turtle might wake up hours or days later.

Stage 4. Medullary paralysis

- Bloodstream → heart → cerebral cortex → cerebrum → cerebellum
- Depresses breathing, heartbeat, blood pressure→anoxia→brain cell death* (~40 sec) → possible heart fibrillation, agonal breaths, muscle spasms (unconscious, reflex)
- Twenty percent of the oxygen consumed by the body is used by the human brain[11] (making ATP to maintain membrane potentials). Thus, the brain is one of the first organs to fail in anoxia, interruption of the oxygen supply to the brain for more than a few minutes leads to irreversible neuronal death.[11]
- Even if a turtle is circulating the pentobarbital quickly enough to progress into Stage 4 and stop the heart, the brain cells do not die within a few minutes as in mammals. Turtle brains are so tolerant to hypoxia (see previous physiology described) that even with the heart stopped, brain cells could still be alive and may be interpreting pain.

Confirmation of Death

Confirmation of death in turtles can be difficult. The AVMA Guidelines state: "it is often difficult to ascertain that an amphibian or reptile is, in fact, dead." Although this statement has become a paradigm when dealing with euthanasia of herpetologic species, the truth is there are methods to ensure reptiles and amphibians are indeed dead, and it is our responsibility as veterinarians to educate ourselves and apply those methods in a manner that respects their welfare.

It seems remedial to discuss how we confirm death; however, with the unique physiology of turtles, it bears some extra attention. Traditionally, many interesting practices are performed to ensure death of turtles. These include the following.

After sodium pentobarbital, leave the turtle on a heating pad for 24 hours

While this method may help the pentobarbital circulate quick enough to bump the turtle from State 3 to Stage 4, it often is not enough to fully stop the heart. Even if it is, the heart stopping doesn't mean the brain cannot perceive pain, because brain cells are so tolerant to hypoxia. The brain could still be interpreting pain.

Check for a heartbeat with a doppler

Turtle hearts can beat so slow and still maintain life. During normal brumation, painted turtle hearts beat a little over 1 beat per 10 min. No veterinarian has placed a doppler probe on a turtle, listened for 10 minutes and then felt confident they were in the correct location enough to declare no heartbeat. Even if that did happen, again, lack of heartbeat does not mean the brain cell death. The brain could still be interpreting pain.

Declare death when the smell of decay is present

The smell of decay only means body cells are dying; it does not mean brain cells are dying. Review physiology above; turtle brains are so tolerant to hypoxia; they can survive past body cells surviving. The brain could still be interpreting pain.

Physically destroying the cerebrum (after anesthesia, sodium pentobarbital)

Pain is perceived in the cerebrum. This is where suffering happens. Death should always be confirmed by physical intervention, that is, destruction of the cerebrum. Death is judged to have occurred when recovery of brain activity is not possible.[7] Humane methods must achieve nonreversible damage or destruction of the brain tissue to ensure death.

Reflexes and other movements of limbs may remain intact after cerebrum destruction because of increased number of spinal reflexes in these species and/or tissues being tolerant to hypoxia. One must ask, if the cerebrum and the ability of the brain to perceive pain is destroyed, does it matter that the limbs are moving, or the heart is beating? The answer is no.

Author's Recommendation for Turtle Euthanasia

The entire sequence below is performed on every turtle euthanasia at WRCMN. It takes ~45-60 minutes time in total but ensures a complete death, no continued perception of pain, and prevents reanimation. After the below procedure, the body can immediately go into the freezer or be destroyed.

1. Alfaxalone double surgery dose (10-20 mg/kg) SC or IM into forelimbs. Subcutaneous is less reliable because may be inadvertently administered into a fat pad but intramuscular may be painful because of the large volume needed. Consider administering partial amounts into both forelimbs, or partial amount subcutaneously. In the author's experience, forelimb administration is more reliable vs rear limb administration, presumably due to the renal-portal system. Wait 20 minutes before proceeding to next step. If turtle is not anesthetized in 20 minutes, repeat this step.

2. Propofol 10-20 mg/kg IV. If the turtle is too small to hit a vein, utilize the occipital venous sinus. Wait 20 minutes before proceeding to next step.

3. Pentobarbital IV same volume as propofol. The occipital venous sinus is the most convenient venous access for this step in the author's experience. Wait 20 minutes before proceeding to next step. After this step, painted turtles' neck will extend dorsally; snapping turtles do not.

4. Destroy the cerebrum. The author prefers to destroy the cerebrum by pithing it with a long needle/pin. Measure from the back of the occipital protuberance to the eyes and ensure your needle/pin is long enough. Insert the needle/pin as you would for an occipital venous stick but advance it further. Scrape the inside of the cranium to confirm you are in the correct location and move the needle/pin around to destroy the cerebrum. Remember you need to destroy the cerebrum, not just the brainstem. Other physical methods may be used instead of pithing but are outside the scope of this class.

References

1. Canadian Council on Animal Care. Overview of acceptable euthanasia methods. *Canadian Council on Animal Care Guidelines on: Euthanasia of Animals Used in Science.* 2010:12-16.

2. Leary S, Underwood W, Anthony R, et al. *AVMA Guidelines for the Euthanasia of Animals: 2020 Edition.* 2020.

3. Meyer RE. Euthanasia and humane killing. In: Grimm KA, Lamont LA, Tranquilli WJ, et al, eds. *Veterinary Anesthesia and Analgesia.* 5th Ed. Ames, IA: Wiley Blackwell; 2015:130-146.

4. Hawkins P, Prescott MJ, Carbone L, et al. A good death? Report of the second Newcastle meeting on laboratory animal euthanasia. *Animals.* 2016;6(9):1-28. doi:10.3390/ani6090050.

5. Ultsch GR. The ecology of overwintering among turtles: where turtles overwinter and its consequences. *Biol Rev Camb Philos Soc.* 2006;81(May):339-367. doi:10.1017/S1464793106007032.

6. Jackson DC, Ultsch GR. Physiology of hibernation under the ice by turtles and frogs. *J Exp Zool. Part A Ecol Genet Physiol.* 2010;313 A(6):311-327. doi:10.1002/jez.603.

7. Pizzi R, McArthur S. Euthanasia technique for chelonians. *Vet Rec.* 2004;154(19):607-608.

8. Krivoruchko A, Storey KB. Turtle anoxia tolerance: Biochemistry and gene regulation. *Biochim Biophys Acta.* 2015;1850(6):1188-1196. doi:10.1016/j.bbagen.2015.02.001.

9. Nilsson GE, Lutz PL. Anoxia tolerant brains. *J Cereb Blood Flow Metab.* 2004;24(5):475-486. doi:10.1097/00004647-200405000-00001.

10. Kesaraju S, Schmidt-Kastner R, Prentice HM, Milton SL. Modulation of stress proteins and apoptotic regulators in the anoxia tolerant turtle brain. *J Neurochem.* 2009;109(5):1413-1426. doi:10.1111/j.1471-4159.2009.06068.x.

11. Larson J, Drew KL, Folkow LP, Milton SL, Park TJ. No oxygen? No problem! Intrinsic brain tolerance to hypoxia in vertebrates. *J Exp Biol.* 2014;217(Pt 7):1024-1039. doi:10.1242/jeb.085381.

12. Milton SL, Prentice HM. Beyond anoxia: the physiology of metabolic downregulation and recovery in the anoxia-tolerant turtle. *Comp Biochem Physiol A Mol Integr Physiol.* 2007;147(2 SPEC. ISS.):277-290. doi:10.1016/j.cbpa.2006.08.041.

13. Storey JM, Storey KB. Natural freezing survival in animals. *Annu Rev Ecol Syst.* 1996;27:365-386.

14. Reese SA, Stewart ER, Crocker CE, et al. Geographic variation of the physiological response to overwintering in the painted turtle (*Chrysemys picta*). *Physiol Biochem Zool.* 2004;77(4):619-630.

15. Kesaraju S, Milton SL. Preliminary evidence of neuronal regeneration in the anoxia tolerant vertebrate brain. *Exp Neurol.* 2009;215(2):401-403. doi:10.1016/j.expneurol.2008.10.017.

16. Packard GC, Packard MJ. Natural freeze-tolerance in hatchling painted turtles? *Comp Biochem Physiol Part A Mol Integr Physiol.* 2003;134(2):233-246. doi:10.1016/S1095-6433(02)00264-7.

17. Cooper JE. Humane euthanasia and post-mortem examination. In: Girling S, Raiti P, eds. *BSAVA Manual of Reptiles.* 2nd ed. Gloucester, UK: British Small Animal Veterinary Association; 2004:168-183.

18. Swiss Veterinary Office. Expert Panel. *Analysis on Humane Killing Methods for Reptiles in the Skin Trade.* 2013.

19. Tranquilli WJ, Grimm KA. Use, definitions, history, concepts, classification and considerations for anesthesia and analgesia. In: Grimm KA, Lamont LA, Tranquilli WJ, et al, eds. *Veterinary Anesthesia and Analgesia.* 5th Ed. Ames, IA: Wiley Blackwell; 2015:5.

20. Warren K. Reptile euthanasia – no easy solution? *Proc Annu Conf Austr New Zealand Council Care Anim Res Teach Thinking Outside the Cage: A Different Point of View.* 2012;111-113.

Risk Factors Influencing Brumation Success in Captive Tortoises in the United Kingdom

Joanna Hedley, BVM&S, DZooMed (Reptilian), DipECZM (Herpetology), MRCVS,
Vicki Baldrey, BVSc, BSc (Hons), DZooMed (Avian), FHEA, MRCVS,
Louis Guezennec, BVetMed, MRCVS

Session #2563

Affiliation: From the Royal Veterinary College Exotics Service, Royal College Street, London, NW1 0TU, UK.

Note: Abstract format selected by author(s). See Front Matter for description of conference proceedings manuscripts submitted as scientific abstracts.

Abstract: The aim of this study was to evaluate brumation methods being used for pet tortoises in the United Kingdom and to identify potential risk factors for mortality and morbidity. Between March and November 2019, tortoise owners were invited to complete a survey on their approach to brumating their tortoise, including details of signalment, preparation for brumation, method and outcome. The study was limited to the Horsfield's (*Testudo* [*Agrionemys*] *horsfieldii*), Hermann's (*T. hermanni*) and spur-thighed tortoise (*T. graeca*). A total of 252 owners completed the questionnaire, providing information on 270 tortoises. Binary logistic regression and multivariate modelling were used to evaluate the potential predictors associated with brumation mortality and those associated with post-brumation problems. A 7.78% mortality rate was found, with 7.22% of surviving tortoises reported to have post-brumation health concerns. Brumation in a garden and an uncontrolled reduction of temperature prior to brumation were the main risk factors for morbidity and mortality in the study. Veterinarians should be aware of these risk factors to ensure they target husbandry and brumation advice to owners of tortoises to reduce these risks.

Chronic Intermittent Esophageal Prolapse in a Red-Eared Slider Turtle (*Trachemys scripta elegans*)

Kimberly L. Lezama, DVM, Meera Kumar Music, BVM&S, MRCVS,
Anneliese Strunk, DVM, DABVP (Avian Practice),
Alicia McLaughlin, DVM, CertAqV,
Susan B. Fogelson, MS, DVM, PhD, DACVP

Session #2590

Affiliation: From The Center for Bird and Exotic Animal Medicine, Bothell, WA 98011, USA (Lezama, Music, Strunk, McLaughlin), Fishhead Labs, LLC, Pathology Consulting Services, Stuart, FL 34997, USA (Fogelson).

Note: Abstract format selected by author(s). See Front Matter for description of conference proceedings manuscripts submitted as scientific abstracts.

Abstract: A 48-year-old female red-eared slider turtle (*Trachemys scripta elegans*) presented for circumferentially swollen neck and tongue with varying severity for 6-8 months, and glossal swelling for 20 years. Physical exam revealed an irregular mass caudal to an area of glossal erythema and ulceration. Soft tissues in the hyoid region were enlarged. Initial diagnostics were obtained, including complete blood count (CBC), serum chemistry, cytology, and computed tomography (CT) scan. CBC showed signs of basophilia and a regenerative response from prior probable anemia. Cytology was inconclusive, and CT illustrated an ill-defined thickening of the tongue base as well as pneumonia. Chelonian herpesvirus polymerase chain reaction (PCR) was negative. Poor response to antibiotic treatments prompted owner elected endoscopic evaluation and surgical biopsy. Endoscopic exam revealed the presumptive tongue base mass was esophageal prolapse caused by entrapment of the hyoid on the internal plastron. Histopathology of the excised mass showed ulcerative esophagitis and fibrovascular polyp formation, consistent with previous or recurrent trauma. The patient declined despite supportive treatments over the following 2 months and was humanely euthanized. Necropsy and histopathology were performed, which confirmed the presence of pneumonia, glossitis, ulcerative esophagitis, and esophageal spindle cell proliferation. Microscopic features of the esophageal lesions supported the previous biopsy diagnosis of chronic inflammation and trauma. At present writing, this is the first documented case of antemortem diagnosed esophageal prolapse in a chelonian causing clinical signs. This case illustrates that esophageal prolapse should be on the differential list for veterinary clinicians seeing chelonians with oropharyngeal pathologies.[1-17]

References

1. Acar N, Acar T, Cengiz F, Şuataman B, Tavusbay C, Haciyanli M. Endoscopic resection for giant oesophageal fibrovascular polyp. *Ann R Coll Surg Engl.* 2020;102(4):e89-e90.

2. Bindaco ALS, Junior AC, Almeida IC, et al. Isolation and characterization of the aerobic bacterial microbiota of the esophagus and its probable association with obstructive caseous lesions in green turtles (*Chelonia mydas*). *Pesq Vet Bras.* 2020;40(11):922-932.

3. Carpenter JW. *Exotic Animal Formulary.* 4th Ed. St. Louis, MO: Elsevier Saunders; 2013.

4. Del-Angel-Caraza J, Alvarez-Contreras PC, Barbosa-Mireles MC, Quijano-Hernandez IA. Intermittent gastroesophageal intussusception in a kitten resolved with a percutaneous gastropexy. *Acta Veterinaria.* 2016;66(3):413-421.

5. Divers SJ, Stahl SJ. *Mader's Reptile and Amphibian Medicine and Surgery.* 3rd Ed. St. Louis, MO: Elsevier Saunders; 2019.

6. Fossum TW. *Small Animal Surgery.* 5th Ed. St. Louis, MO: Elsevier Saunders; 2019.

7. Gandar F, Marlier D, Vanderplasschen A. In vitro and in vivo assessment of eprociclovir as antiviral treatment against testudinid herpesvirus 3 in Hermann's tortoise (*Testudo hermanni*). *Res Vet Sci.* 2019;124:20-23.

8. Garcia-Morante B, Penzes JJ, Costa T, et al. Hyperplastic stomatitis and esophagitis in a tortoise (*Testudo graeca*) associated with an adenovirus infection. *J Vet Diagn Invest.* 2016;28(5):579-583.

9. Houck EL, Cohen EB, Womble M, et al. Radiographic anatomy and barium sulfate contrast study of the gastrointestinal tract of Eastern box turtles (*Terrapene carolina carolina*). *Vet Radiol Ultrasound.* 2019;60(5):473-484.

10. Jerdy H, Werneck M, Veloso R, et al. Fatal *Rameshwarotrema uterocrescens* infection with ulcerative esophagitis and intravascular dissemination in green turtles. *Int J Parasitol Parasites Wildl.* 2019;9:281-284.

11. Kumar A, Behari A, Kapoor VK, et al. Postendoscopic intraoral prolapse of a giant esophageal fibrovascular polyp. *Formosan J Surg.* 2016;49(4):165-168.

12. Lee S, Park S, Kim M, et al. Transhiatal esophagogastric anastomosis and postoperative monitoring of thoracic esophageal leiomyosarcoma in a dog. *Can Vet J.* 2020;61(4):401-406.

13. Murphy LA, Nakamura RK, Miller JM. Surgical correction of gastro-oesophageal intussusception with bilateral incisional gastropexy in three dogs. *J Small Anim Pract.* 2015;56(10):630-632.

14. Sutton JS, Culp WTN, Scotti K. Perioperative morbidity and outcome of esophageal surgery in dogs and cats: 72 cases (1993-2013). *J Am Vet Med Assoc.* 2016;249(7):787-793.

15. Takeuchi R, Takagi S, Hosoya H, et al. Postoperative improvement of megaesophagus and esophageal motility in a 19-day-old puppy with gastroesophageal intussusception. *J Vet Med Sci.* 2021;83(11):1740-1744.

16. Tecilla M, Bielli M, Origgi FC. Polycystic kidney-like disease in a red-ear slider turtle (*Trachemys scripta elegans*). *J Comp Pathol.* 2018;164:44-47.

17. Włodarczyk J, Warmus J, Gocyk W, Kużdżał J. Giant oesophageal polyp prolapse. *Eur J Cardiothorac Surg.* 2013;43(6):e198.

Critical Care of Pancake Tortoises (*Malacochersus tornieri*): A Rescue

Cathy A. Johnson-Delaney, DVM

Session #2672

Affiliation: From Northwest Zoological Supply, 3614 Smith Ave, Everett, WA 98034, USA.

Note: Abstract format selected by author(s). See Front Matter for description of conference proceedings manuscripts submitted as scientific abstracts.

Abstract: Twenty-four adult pancake tortoises (*Malacochersus tornieri*) were imported from Europe to start a preservation breeding colony. Transit was over 24 hours and all arrived clinically dehydrated with body temperatures below 18.5 °C (65 °F). The CITES documentation listed 4 males and 20 females; however, there were 3 males and 21 females. All were deemed clinically emaciated and all were placed into warmed electrolyte solution for hydration correction upon unpacking. The tortoises were grouped by weight, with one male in three of the habitats. Habitats contained shelters, UV-B lighting, heat emitters and water dishes, details will be discussed in the lecture. Over the next few days, care included intensive supportive care (fluids, parenteral vitamin B complex, calcium gluconate, tube feeding, etc.). Fecal parasite examination revealed a heavy load of nematodes, which may be commensal in herbivorous species. Due to the debilitated condition, anthelminthic treatment was not used, rather appropriate diet and husbandry were provided, with repeated examinations showing decreases in the parasite load. One tortoise died. The remaining 23 began eating within 3-5 days of arrival. The diet was made fresh daily, supplemented with calcium. All tortoises were weighed weekly and any not gaining weight were administered supportive care as listed above. After 6 weeks the fecal parasite examinations were negative. After 7 months under daily veterinary care, the breeding colony was tested for intranuclear coccidiosis and found negative. The natural history, biology captive diet, husbandry and veterinary care of this species are discussed.[1-7]

References

1. Morris, P. Pancake tortoise (*Malacochersus tornieri*). World Chelonian Trust care sheet. Available at http://www.chelonia.org/Articles/Mtorniericare.htm/ Accessed March 15, 2022.

2. Eustace A, Esser LF, Mremi R, et al. Protected areas network is not adequate to protect a critically endangered East Africa chelonian: modelling distribution of pancake tortoise, *Malacochersus tornieri* under current and future climates. *PLoS ONE.* 2021;16(1):e0238669. https://doi.org/10.1371/journal.pone.0238669

3. Mautner AK, Latimer AE, Fritz U, Scheyer TM. An updated description of the osteology of the pancake tortoise *Malacochersus tornieri* (Testudines:Testudinidae) with special focus on intraspecific variation. *J Morphol.* 2017;278(3):321-333.

4. Griffin C, Reavill DR, Stacy BA, et al. Cryptosporidiosis caused by two distinct species in Russian tortoises and a pancake tortoise. *Vet Parasitol.* 2010;170(1-2):14-19.

5. Traversa D. Evidence for a new species of *Cryptosporidium* infecting tortoises: *Cryptosporidium ducismarci*. *Parasit Vectors.* 2010;25:21.

6. Ewert MA, Hatacher, RE, Goode JM. Sex determination and ontogeny in *Malacochersus tornieri*, the pancake tortoise. *J Herpetol.* 2004;38(2):291-295.

7. Malonza PK. Ecology and distribution of the pancake tortoise, *Malacochersus tornieri* in Kenya. *J East Afr Nat Hist.* 2003;92(1):81-96.

Exophthalmia Associated with Atherosclerosis and Myocardial Fibrosis in a Bearded Dragon (*Pogona vitticeps*)

Marianne Caron, DVM, and
Thomas Boyer, DVM, DABVP (Reptile and Amphibian Practice)

Session #2683

Affiliation: From Pet Hospital of Penasquitos, 9888 Carmel Mountain Rd Ste F, San Diego, CA 92129, USA.

Note: Abstract format selected by author(s). See Front Matter for description of conference proceedings manuscripts submitted as scientific abstracts.

Abstract: Exophthalmia, a common clinical sign in bearded dragons, was observed with severe aortic atheromatous plaque and myocardial fibrosis in an adult bearded dragon. An 8-year-old intact male bearded dragon presented for anorexia, lethargy, and lack of defecation. The patient had a history of obesity, with recent weight loss. Bilateral exophthalmia with caudal displacement of the globes and pulsing was noted in the caudodorsal aspect of the oral cavity. The thoracic limbs were subjectively mildly swollen. Radiographs demonstrated an enlarged heart compared to radiographs taken approximately 2 months prior. Cardiac ultrasound revealed thickened ventricular walls, hyperechogenicity of the myocardium, and mild free fluid around the liver. A tentative diagnosis of cardiac failure was given. The owner elected euthanasia without further workup or treatment 2 days later. Necropsy confirmed thickened ventricular walls with decreased luminal space, a large circumferential yellow deposit within the aortic wall, and a pale liver. Histopathology found aortic atherosclerosis with suspected secondary myocardial fibrosis, as well as marked hepatocellular lipidosis progressing towards hepatic fibrosis and mild biliary hyperplasia. This case demonstrates some clinical signs associated with atherosclerosis (exophthalmia, swollen thoracic limbs, and pulsing in the caudodorsal oral cavity) not previously described in bearded dragons.[1,2]

References

1. Schilliger L, Lemberger K, Chai N, et al. Atherosclerosis associated with pericardial effusion in a central bearded dragon (*Pogona vitticeps*). *J Vet Diagn Invest.* 2010;22(5):789-792.

2. Schilliger L, Paillusseau C, Gandar F, et al. Hypertensive heart disease and encephalopathy in a central bearded dragon (*Pogona vitticeps*) with severe atherosclerosis and first-degree atrioventricular block. *J Zoo Wildl Med.* 2019;50(2):482-486.

Intravenous Catheterization in Client-Owned Lizards at a Veterinary Teaching Hospital: 21 Cases (2018-2021)

Marina Liles, DVM,
João Brandão, LMV, MS, DipECZM (Avian),
Nicola Di Girolamo, DMV, MSc (EBHC), PhD, DipECZM (Herpetology), Dipl ACZM

Session #2731

Affiliation: From Department of Veterinary Clinical Sciences, College of Veterinary Medicine, Oklahoma State University, Stillwater, OK 74078, USA.

Note: Abstract format selected by author(s). See Front Matter for description of conference proceedings manuscripts submitted as scientific abstracts.

Abstract: Intravenous (IV) catheter placement is one of the most common medical procedures performed in human and veterinary medicine. Complications in patients with an IV catheterization include phlebitis, infection, obstruction, leakage, and accidental removal of catheters. Placement of IV catheters in exotic animal medicine may be more challenging, especially in reptiles. There is limited knowledge of what catheter sites and characteristics are used more often in reptiles in a clinical setting. The purpose of this study is to assess the prevalence of lizards presenting to a veterinary teaching hospital that had an intravenous catheter placed, the catheterization sites used, and complications arising with the placement of the catheter both in the short- and long-term. Of 129 lizard consultations performed between September 27, 2018 to September 27, 2021, 21 resulted in a catheter placed, resulting in an overall prevalence of 16.3% catheters placed. Reason for catheter placement included hospitalization for fluid therapy (10/21), anesthesia and/or surgery ± hospitalization (7/21), computed tomography scan with contrast (2/21), euthanasia only (1/21), and CPR only (1/21). All catheters were placed in the ventral coccygeal vein via a ventral approach. Sedation was used in 6/21 of the catheters placed. Seven of the catheters were used for fluids administration only, four were used for administering drugs/medications only, and nine were given both fluids and drugs/medications. Based on retrospective evaluation, intravenous catheter placement in lizards is feasible in conscious animals, with roughly one consultation out of six resulting in a catheter placed.

Section 35

Hybrid: Mixed Topics

Kimberly Boykin, DVM, MS;
Caitlin Cunningham, DVM;
Lauren Gawel, DVM;
Kaitlyn Nikirk, DVM Candidate (Class of 2024)
Session Chairs

On-Call Exotic Emergency Services: One-Year Retrospective Caseload Analysis

Anna-Marie Ruoff, DVM,
Elizabeth Hyde, DVM,
Jay Johnson, DVM

Session #2534

Affiliation: From Arizona Exotic Animal Hospital, Mesa, AZ 85201, USA (Ruoff, Hyde, Johnson), The Center for Bird and Exotic Animal Medicine, Bothell, WA 98011, USA (Hyde).

Abstract: Exotic animal owners want options for emergency care for their pets, but few 24-hour small animal emergency hospitals are willing or able to handle exotic animal emergencies. We will discuss caseload and outcome data from the first year of an on-call, after-hours, exotic pet emergency service recently established by a high-volume, exotics-exclusive hospital in a large metropolitan area. 624 after-hours calls were fielded over 361 nights, averaging 1.73 calls per night (0-11 calls). Established clients accounted for 28.2% of calls, non-clients 64.1%, unknown client status 6.1%, and consults from other veterinarians 1.6%. Calls referencing small companion mammal patients encompassed 51.4% of calls, birds 18.9%, reptiles 14.9%, large mammals 4.3%, native wildlife 1.6%, amphibians 0.3%, fish 0.3%, and unknown species 8.2%. An after-hours emergency appointment was recommended by the on-call doctor for 59.0% of calls; of those, 34.8% came in for the appointment, 63.3% declined the appointment, and 1.9% were deceased on arrival. There was no significant difference between established clients and non-clients for compliance in making an after-hours appointment when one was recommended (42.1% and 32.9%, respectively). This information may guide other exotic animal practitioners to evaluate potential caseload, outcomes, and viability of establishing their own after-hours emergency services.

Introduction

Arizona Exotic Animal Hospital (AZEAH) is an exotic animal exclusive hospital with offices in Mesa, Phoenix, and Tucson, AZ. In October 2019, AZEAH established an after-hours on-call emergency service for its clients in the greater Phoenix metropolitan area. People who called the hospital after business hours were given the option to contact an on-call doctor if their pet was experiencing a medical emergency. These self-selected phone calls were forwarded through the Jive mobile VoIP application to the on-call doctor's cellular phone. Based on the signalment, presenting complaint, medical history, and/or client request, the on-call doctor could recommend and arrange for an after-hours emergency appointment to deliver care. Data from the phone calls were collected to better evaluate the needs of clients and the hospital. Data collected included date, species of the pet, presenting complaint and history, whether the owner was an established client of the hospital, the on-call doctor's recommendations, and whether the call resulted in an after-hours emergency appointment.

Caseload Analysis

In its first year, the after-hours emergency service received 624 calls over 361 nights. Data were not recorded for three nights (0.82% of 364 nights) due to recorder error or application malfunction; these were excluded from analysis. The average call volume was 1.73 calls per night, ranging from 0 calls to 11 calls. No calls were received on 109 nights (30.2%).

At the initiation of the service, business hours ended at 1800 hours (6:00 pm). After 7 months of the on-call service, the data were internally evaluated. The business hours of the hospital changed to accommodate the heavy call load immediately after closing; for the remaining 133 nights of analysis, business hours ended at 2200 hours (10:00 pm).

There were 481 calls during the 231 nights before the business hour change. Data were not recorded for three nights (1.3% of 231) due to recorder error or application malfunction; these were excluded from further analysis. The average call volume was 2.11 calls per night. No calls were received on 49 nights (21.5%).

There were 143 calls during the 133 nights after the business hour change. The average call volume was 1.08 calls per night. No calls were received on 60 nights (45.1%). The number of nights without calls differed significantly before and after the business hour change, X^2 (1, n = 361) = 22.24, $p < 0.05$.

Of the 624 total calls, 176 were from established clients (28.2%), 400 were from non-clients (64.1%), 10 were consults from other veterinarians (1.6%), and 38 were unknown status (6.1%, typically due to an unreturned voicemail message). Before the hours change, 140 were from established clients (29.1%), 315 were from non-clients (65.5%), 7 were consults from other veterinarians (1.5%), and 19 were unknown status (4.0%). After the hours change, 36 were from established clients (25.2%), 85 were from non-clients (59.4%), 3 were consults from other veterinarians (2.1%), and 19 were unknown status (13.2%). The clientele utilizing the service differed significantly before and after the business hour change, X^2 (3, n = 624) = 17.30, $p < 0.05$.

Clients were asked what species their pet was during most phone calls; however, the level of client specificity varied (e.g., jenday conure versus bird). The authors chose to categorize species data to allow for descriptive analysis. The categories were small companion mammals (including rabbits, guinea pigs, rats, hedgehogs, ferrets, sugar gliders, etc.), large mammals (including miniature pigs, small ruminants, camelids, macropods, etc.), birds, reptiles, amphibians, fish, native wildlife (which were excluded from previous categories), and unknown species (typically from unreturned voicemail messages). Calls referencing small companion mammal patients encompassed 51.4% of calls, birds 18.9%, reptiles 14.9%, large mammals 4.3%, native wildlife 1.6%, amphibians 0.3%, fish 0.3%, and unknown species 8.2%.

The resolution of each call was recorded by the on-call doctor who received it. These resolutions were then categorized for descriptive analysis. The on-call doctor recommended an after-hours emergency appointment for 368 calls (59.0%). Recommendations to monitor the pet's condition and/or make a daytime appointment were made during 167 calls (26.8%). For clinically unstable patients such as those in acute respiratory distress, the on-call doctor advised the caller to take the patient to the nearest available emergency facility (21 calls, 3.4%). Occasionally, the owner called while their pet was being treated at another veterinary facility or had already elected to go to another veterinary facility (19 calls, 3.0%). During 12 calls (1.9%), the pet had already died; condolences, necropsy, and cremation services were offered. Calls about native wildlife were referred to a local wildlife rehabilitation organization (10 calls, 1.6%). Veterinarians from other hospitals made 10 calls (1.6%) for consultation about exotic animal patients. Pets too large to transport to the hospital for care were advised to contact large animal mobile veterinarians as emergency house calls were not offered by this service (6 calls, 1.0%). Five callers (0.8%) were calling from out of state and recommended to consult a local veterinarian. Calls about laboratory results, hospital hours, or other non-patient-related questions were requested to call back during regular business hours (6 calls, 1.0%).

Of the 368 calls for which an after-hours emergency appointment was recommended, 128 resulted in an appointment (34.8%), 233 declined the appointment (63.3%), and 7 were deceased on arrival (1.9%).

Established clients were advised to come in for an after-hours emergency appointment for 102 of 176 calls (58.0%). Of those, 43 came in for an appointment (42.1%), 56 declined an appointment (54.9%), and 3 pets were deceased on arrival (2.9%). Non-clients were advised to come in for an after-hours emergency appointment for 258 of 400 calls (64.5%). Of those, 85 came in for an appointment (32.9%), 169 declined an appointment (65.5%), and 4 pets were deceased on arrival (1.6%). There were 38 calls from people of unknown client status, but these were typically short calls without much patient information, blocked numbers that did not leave identifying information for a return call, or a caller not answering or returning the doctor's call(s). An after-hours emergency visit was recommended in 8 (21%) of these cases, typically based on the severity of the presenting complaint, but no visits occurred (0%).

The authors hypothesized that established clients would be more likely to heed the on-call doctor's recommendation for an after-hours emergency appointment than non-clients; however, there was no significant difference between established clients and non-clients for compliance with recommended after-hours emergency appointments, X^2 (1, n = 353) = 3.06, p = 0.08. Phone calls with unknown client status were excluded from this analysis.

Conclusions

During the first year of the after-hours on-call service, 624 phone calls resulted in 128 after-hours emergency appointments. The service benefited both established clients and non-clients. Established clients utilized the service frequently and were able to get advice about or care for their pet during extended business hours or after-hours emergency appointments. Each phone call with a non-client was a point of contact for a potential new client, including if they came in for an after-hours emergency appointment or chose to schedule a daytime appointment. Other benefits to the practice included maintaining relationships with local veterinarians by offering after-hours consultations for exotic pet emergencies.

The service has progressed beyond its first year with continued success. In addition to the business hours change previously mentioned, other changes to staffing, on-call doctor scheduling, after-hours call record-keeping, and on-call software were made based on call data analysis and on-call doctor feedback. Further analysis, including continued call data collection and financial analysis will continue to guide business decisions regarding the service. The authors hope that these data can also serve to guide the creation of after-hours on-call emergency services at other veterinary facilities to better serve exotic pets and their owners.

Acknowledgments: The authors acknowledge the veterinarians, veterinary technicians, and paraprofessional staff of Arizona Exotic Animal Hospital for their support through the first year of the on-call emergency service.

Pandemic-Era Exotic Medicine: A Review of Changes in Medical Practices, Infrastructure, and Technology to Accommodate Unprecedented Circumstances

Alicia McLaughlin, DVM, CertAqVet, and
Meera Music, BVM&S, MRCVS

Session #2593

Affiliation: From The Center for Bird and Exotic Animal Medicine, 11401 NE 195th St Bothell, WA 98011, USA.

Abstract: The worldwide SARS-CoV-2 (COVID-19) pandemic that began affecting the United States in early 2020 resulted in major uncertainty and upheaval across society, including in veterinary clinics. The Center for Bird and Exotic Animal Medicine (CBEAM) is an exotics-exclusive hospital in Bothell, WA, that is staffed 24/7 and provides both routine and emergency (ER)/critical care to exotic pets. Since the start of the pandemic, the hospital experienced a drastic (50%) increase in patient numbers and high demand for emergency services, including a tripled average daily emergency caseload. This unexpectedly rapid growth resulted in a variety of necessary changes to patient intake procedures, hospital infrastructure, and technology to accommodate the needs of staff members and the community. The authors hope that sharing their experiences will help other clinics dealing with similar challenges.

Pandemic Medicine

The worldwide SARS-CoV-2 (COVID-19) pandemic that began impacting the United States in early 2020 resulted in major uncertainty and upheaval across society, including in veterinary clinics. To follow recommendations to protect staff and continue to operate, clinics had to rapidly change how they practiced medicine (with protocols changing as often as week-to-week to meet recommendations from the Centers for Disease Control and Prevention). Many practices switched to curbside medicine, where clients were not permitted to enter the veterinary hospital with their pets, to minimize exposure time between staff and clients. This new appointment structure resulted in decreased efficiency, decreased appointment availability, and longer wait times for clients. It also posed an additional challenge to build rapport with new clients due to a lack of face-to-face interaction with clients.

In addition to the significant biosecurity concerns regarding exposure of staff to clients and to other staff members, many issues arose surrounding use of biomedical equipment, including surgical gloves and masks. There were shortages of equipment (e.g., gloves, masks, medications) and mixed public health messages surrounding the benefits of biosecurity measures. In some cases, government mandates impacted how hospitals could operate (e.g., limiting routine surgical procedures, for example).

Staff availability was significantly affected by changes in childcare availability, changes in procedures connected to staff illness (e.g., quarantine measures, etc.), concerns about personal health, and burnout.

These issues were further compounded by a huge boom in new pet ownership.[1] People developed stronger bonds with their pets and noticed changes in behavior and medical conditions that might have been missed in the past due to increased time at home during lockdowns. The limited numbers of veterinarians graduating from college was insufficient to meet this increased demand for care. In addition, veterinarians were prone to the same life challenges everyone else experienced during the pandemic, including vulnerable family members, unexpected deaths, and changing access to childcare.

The decreased access to veterinary services associated with all of these factors was very stressful to both veterinarians and to pet owners. Medical care was impacted for many animals, sometimes to the point of loss of life. There was also a significant increase in inappropriate behavior of clients towards veterinary staff, including both physical and verbal threats and online campaigns against individual clinics. Veterinary staff were overworked and exhausted, increasing the potential for mistakes to occur (either communication or medical), and burnout in the field was rampant. Emergency veterinary clinics were utilized inappropriately for non-urgent medical concerns due to lack of routine appointment availability, leading to even longer wait times and frustration on both the part of staff and clients. Preventative care was delayed for many animals, which led to development of other illnesses.

One Clinic's Journey

The Center for Bird and Exotic Animal Medicine (CBEAM) is an exotics-exclusive hospital in Bothell, WA, that is staffed 24/7 and provides both routine and emergency/critical care to exotic pets. CBEAM is the only exotics-exclusive hospital with round-the-clock emergency (ER) services and staffing in the Pacific Northwest at this time. Since the start of the pandemic, the hospital experienced a drastic (50%) increase in patient numbers and high demand for emergency services, including a tripled average daily emergency caseload. This unexpectedly rapid growth resulted in the hospital leadership team making a variety of necessary changes to patient intake procedures, hospital infrastructure, and technology to accommodate the needs of staff members and the community.

Triage system

One of the hardest changes for CBEAM staff to make was to accept the fact that they could not continue to provide care for every single client that sought veterinary care for their exotic pet. The hospital did not have enough staffing at the start of the pandemic to accommodate the number of emergency patients demanding services on top of hospitalized, critically ill patients, and emergency surgeries, much less routine wellness care. The clinic significantly decreased appointment availability for preventative care to be able to redirect veterinarians as emergency clinicians while actively hiring additional clinicians and support staff to accommodate the extreme growth curve.

Even with this shift in resources, the ER service would routinely be overwhelmed. As a result, the clinic took a multi-pronged approach to ER scheduling.

1. Walk-ins were actively discouraged, though if a walk- in occurred, triage would occur on-site, and wait times would be appropriately estimated.

2. Emergency patient intake numbers were limited to three per hour, permitting 20 minutes per emergency patient for examination, stabilization treatments, and communication with clients to establish a plan for the day.

3. A color-coded triage system was implemented to guide emergency patient scheduling. If the emergency service was overwhelmed, a color code for which level of emergency patient would be accepted that day was initiated.

4. Specific time frames within the schedule were blocked off to prevent further emergencies from arriving during these times to permit care of hospitalized patients and diagnostic workups/procedures of previously admitted patients.

The goal for these policies was to prevent staff burnout and decrease the risk of medical errors while improving patient care for the hospitalized pets.

Space utilization

Boarding services were discontinued to allow staff to focus on caring for sick pets, and this facility space was repurposed. Previously used office space was transitioned into a triage area for ER patients. The leadership team developed new services and restructured others to mitigate the space limitations of the parking lot. The goal was to minimize the amount of time clients were present at the clinic to free up parking space for incoming ER clients. ER patients were dropped off shortly after arrival (assuming the client was reachable by phone). A new urgent care service helped accommodate scheduling of less critical emergencies to prevent the ER service from being overwhelmed. Rechecks of sick patients were scheduled as drop off appointments.

Staffing and scheduling changes

The leadership team increased support staff wages across the board to minimize turnover and address the needs of a rapidly changing economy. Priority was placed on hiring additional veterinarians and support staff as soon as possible to address the needs of the hospital. Members of the team that had not previously served in a leadership capacity were given new responsibilities to help improve cohesiveness of a rapidly growing team. A new training program was developed, since the old method of training staff on-the-job was not feasible with the fast pace of the clinic floor. A new staff shift (swing shift) was added to improve functionality of shift work and to prevent staff from feeling the need to work overtime to meet the needs of patients, further reducing incidence of burnout.

Technology upgrades

New technology was added to the hospital to address the shortcomings of client communication while practicing curbside medicine, including the use of a texting service (AllyDVM), a new interdepartmental communication process (Avimark Whiteboard), automated forms for history taking (Jotforms), and increased phone lines and updated computer systems. A new Client Communicator role was created to facilitate communication between the clients, reception team, and medical team, ultimately improving the client experience.

Client management

A no-tolerance policy for client abuse was instituted.

Summary

There is a new coalition being put together by Penn Vet to address the national crisis in pet care caused by the pandemic.[2] It will be interesting to see what solutions the coalition comes up with, and how veterinary clinics continue to evolve to meet the needs of their communities. The presenters hope that the description of their changing procedures to meet their community's needs will be helpful to other veterinarians facing similar challenges.

Acknowledgments: The authors thank Team CBEAM for their ongoing support, hard work, and dedication to provide the best care possible for exotic pets.

References

1. American Veterinary Medical Association. Pet population still on the rise, with fewer pets per household. Available at: https://www.avma.org/javma-news/2021-12-01/pet-population-still-rise-fewer-pets-household. Accessed March 1, 2022.

2. Columbia Broadcasting System. New Coalition Put Together By Penn Vet Aiming To Improve 'National Crisis' in Pet Care Caused By Pandemic. Available at: https://philadelphia.cbslocal.com/2021/09/23/penn-vet-coalition-pet-care-national-crisis-pandemic/ Accessed March 1, 2022.

The Impact of Dermal Characteristics on Low-Level Laser Power Measurement in Post-Mortem Zoological Species

Faith Burgos, DVM,
Michelle Borsdorf, DVM, Dipl ACZM,
John Ladner, LVT, MLSO,
Anne White, DVM, PhD, CVA,
Tara M. Harrison, DVM, MPVM, Dipl ACZM, Dipl ACVPM, DipECZM (Zoo Health Management), CVA

Session #2585

Affiliation: From North Carolina State University College of Veterinary Medicine, Department of Clinical Sciences, Raleigh, NC 27607, USA, Exotic Species Cancer Research Alliance, North Carolina State University College of Veterinary Medicine, Department of Clinical Sciences, Raleigh, NC 27607, USA (Harrison).

Note: Abstract format selected by author(s). See Front Matter for description of conference proceedings manuscripts submitted as scientific abstracts.

Abstract: The purpose of this study was to evaluate the power of a low-level laser after penetration through dermal tissue (~1 mm thickness) using a variety of post-mortem animal tissue with an ActivetPro low level laser similar to a previous study done in rats.[1] Specifically, this study evaluated the impact of fur, feathers, scales, and different pigments on penetration of the laser light. Frozen and thawed skin tissue samples were placed inside a light restricted box and exposed to a pre-programmed laser level from a Multiradiance ActivetPro photobiomudulation (PMB) device, and measured with a power meter to evaluate the light penetration through the tissue samples. Light penetration measurements were recorded at seven timepoints (range, 1-150 seconds). Notably, integument color appeared to impact power readings. In rabbits (*Oryctolagus cuniculus*), white hair had the highest power readings (14.98 Hz), followed by fawn/grey colored (0.83 Hz), then black colored (-0.05 Hz). Similarly in iguanas (*Iguana iguana*) and bullfrogs (*Lithobates catesbeianus*), lighter colored/pigmented specimens had higher power readings than darker specimens (16.42 Hz, 20.56 Hz respectively). In sea turtles, Kemp's Ridley (*Lepidochelys kempii*) had higher power readings than green sea turtles (*Chelonia mydas*) (36.88 Hz, 23.16 Hz respectively). In hawks (*Buteo jamaicensis*), covert and down feathers on the legs inhibited laser power measurements to minimal to none (3.15 Hz); however, when the feathers were damp, which allowed the skin tissue to have direct contact with the laser, power readings were measured (21.14 Hz). This is the first study, to the authors' knowledge, that investigates the effectiveness of low-level laser penetration through various species' tissues.

References

1. Joensen J, Ovsthus K, Reed RK, et al. Skin penetration time-profiles for continuous 810 nm and superpulsed 904 nm lasers in a rat model. *Photomed Laser Surg.* 2012;30(12):688-694.

2. Archibald KE, Harrison TM, Troan B, et al. Effect of multiradiance low-level laser therapy and topical silver sulfadiazine on healing characteristics of dermal wounds in marine toads (*Rhinella marina*). *Vet Med Int.* 2020;8888328.

3. Shaver SL, Robinson NG, Wright BD, et al. A multimodal approach to management of suspected neuropathic pain in a prairie falcon (*Falco mexicanus*). *J Avian Med Surg.* 2009;23(3):209-213.

4. Hewitt PG. *Conceptual Physics.* 9th Ed. Boston, MA: Addison Wesley; 2001.

Computed Tomography Basics and Artifacts

M. Scott Echols, DVM, DABVP (Avian Practice)

Session #2577

Affiliation: From Echols Veterinary Services, Salt Lake City, UT 84121, USA, Parrish Creek Veterinary Clinic, 86 N 70 W St, Centerville, UT 84014, USA, Medical Center for Birds, 3805 Main St, Oakley, CA 94561, USA.

Note: Abstract format selected by author(s). See Front Matter for description of conference proceedings manuscripts submitted as scientific abstracts.

Abstract: Computed tomography (CT) is becoming commonplace in avian and exotic animal medicine. Over the past 30 years, CT technology has rapidly advanced creating a diverse array of products and capabilities. Simply stated, not all CT scanners are equal or best suited for small avian and exotic animal species commonly encountered in practice. In particular, the newer high-definition volumetric cone beam scanners are providing significantly improved resolution and contrast over the more traditional slice based and older cone beam CT units. Contrast studies can dramatically improve visualization of soft tissue structures and better define normal versus abnormal anatomy not possible with plain CT. Newer iodine-based contrast protocols include higher doses (4-6 ml/kg IV in birds and reptiles and 3-4 ml/kg IV in exotic mammals) and longer administration times (full dose delivered over 2 minutes) compared to previously published protocols (generally consisting of 1-3 ml/kg IV delivered as a rapid bolus). Combined with pre, during and post-contrast CT scans, these new protocols are providing functional and significantly more patient data compared to older protocols. Additionally, CT technology is complex, and artifacts are common which can adversely affect interpretation. Artifacts can be divided into several types including physics based, patient based, scanner based and more. This presentation covers currently available CT technology, newer contrast protocols, and common artifact presentation and resolution.

The Role of the Veterinarian in the Newly Emerging Insects-as-Feed-and-Food Industry

Kimberly L. Boykin, DVM, MS, and
Mark A. Mitchell, DVM, MS, PhD, DipECZM (Herpetology)

Session #2654

Affiliation: From Louisiana State University School of Veterinary Medicine, Skip Bertman Drive, Baton Rouge, LA 70803, USA.

Note: Abstract format selected by author(s). See Front Matter for description of conference proceedings manuscripts submitted as scientific abstracts.

Abstract: Insects require substantially less resources to produce than traditional sources of animal protein and could help to feed an ever-growing human population with less effects on the environment and climate. Over the last decade, the insects-as-feed-and-food industry has rapidly emerged. As it continues to grow, the need for government regulation increases and veterinarians will likely be called upon to assist in assessing animal health, implementing biosecurity measures, and formulating/prescribing treatment protocols similar to what we have seen with the honeybee industry and the institution of veterinary feed directives (VFDs). Veterinarians will need to become knowledgeable regarding the diseases afflicting the feeder insect industry and how best to diagnosis pathogens of concern. Similar to other agricultural markets, high animal densities and management practices put insects at high risk for infectious diseases. In this presentation, we will focus on the cricket industry which has been particularly hard hit from viral diseases, but will also discuss the role that bacterial, fungal, and parasitic diseases could play in overall cricket health as well as the potential for crossover to humans and other animals. If the insects-as-feed-and-food industry is going to be viable, veterinary involvement will be essential to ensure that insects can be used as a safe source of food for all.

Comparison of Next-Generation DNA Sequencing as an Infectious Disease Diagnostic Tool to Traditional Culture and Sensitivity Testing

Janina A. Krumbeck, PhD

Session #2580

Affiliation: From MiDOG LLC, 14762 Bentley Circle, Tustin, CA 92780, USA, Zymo Research Corp., 17062 Murphy Ave, Irvine, CA 92614, USA.

Note: Abstract format selected by author(s). See Front Matter for description of conference proceedings manuscripts submitted as scientific abstracts.

Abstract: Successful infection treatments, require accurate pathogen identification.[1] Culture and sensitivity testing are considered the gold standard for infectious disease diagnosis for both human and pet patients; however, based on DNA/RNA sequencing research, it is estimated that only 1% of the world's microbes are culturable in the laboratory.[2] Especially for exotic patients, culturing of clinical samples has been challenging and diagnosis often relies on PCR testing to overcome culturing limitations. While PCR testing is fast and provides a high degree of sensitivity and specificity, it is limited to the target panel and provides limited clinical value when the panel returns with negative results. Next-generation DNA sequencing (NGS) has the same benefits as PCR testing, but also offers the de novo identification and quantification of all bacteria and fungi present in a sample, including novel pathogen discovery.[3] In this study, clinical samples from 30 exotic pet patients pre-diagnosed with bacterial and/or fungal infections are tested both with an NGS-based diagnostic tool provided by a commercial NGS-testing lab[4-10] and traditional culture and sensitivity testing. The results are compared for the microorganisms detected, cell numbers, and antimicrobial resistance profile(s). No treatment decisions were considered for this study. Preliminary results highlight a large bacterial and fungal species diversity in the study cohort and showcased the lack of sensitivity of culture testing. Specifically, fungal pathogens and anaerobic bacteria were often missed by culture. NGS has outstanding advantages in the diagnosis of bacterial and fungal infections and optimizes the treatment of infectious diseases.

References

1. Koch L. Metagenomics: pathogen diagnostics for the masses. *Nat Rev Genet.* 2016;17(7):378.

2. Barcina I, Arana I. The viable but nonculturable phenotype: a crossroads in the life-cycle of non-differentiating bacteria? *Rev Environ Sci Biotechnol.* 2009;8:254-255.

3. Tang S, Prem A, Tjokrosurjo J, et al. The canine skin and ear microbiome: a comprehensive survey of pathogens implicated in canine skin and ear infections using a novel next-generation-sequencing-based assay. *Vet Microbiol.* 2020;247:108764.

4. Melgarejo T, Oakley BB, Krumbeck JA, et al. Assessment of bacterial and fungal populations in urine from clinically healthy dogs using next-generation sequencing. *J Vet Intern Med.* 2021;35(3):1416-1426.

5. Krumbeck JA, Reiter AM, Pohl JB, et al. Characterization of oral microbiota in cats: novel insights on the potential role of fungi in feline chronic gingivostomatitis. *Pathogens.* 2021;10(7):904.

6. Niemiec BA, Gawor J, Tang S, et al. The bacteriome of the oral cavity in healthy dogs and dogs with periodontal disease. *Am J Vet Res.* 2022;83(1):50-58.

7. Rexo A, Hansen B, Clarsund M, et al. Effect of topical medication on the nasomaxillary skin-fold microbiome in French bulldogs. *Vet Dermatol.* 2022;33(1):10-e5.

8. Tsang W, Linde A, Krumbeck JA, et al. Occurrence of antimicrobial resistance genes in the oral cavity of cats with chronic gingivostomatitis. *Animals.* 2021;11(12):3589.

9. LafeberVet. The importance of next-generation sequencing in avian veterinary medicine. Available at: https://lafeber.com/vet/the-importance-of-next-generation-sequencing-in-avian-veterinary-medicine/. Accessed May 27, 2022.

10. Niemiec BA, Gawor J, Tang S, et al. The mycobiome of the oral cavity in healthy dogs and dogs with periodontal disease. *Am J Vet Res.* 2021;83(1):42-49.

Hematology and Biochemistry Reference Interval Dynamics in Captive Cockatoos and Bearded Dragons

Thanh Le

Session #2648

Affiliation: From Moichor Inc., 555 De Haro St Suite 220, San Francisco, CA 94107, USA.

Abstract: Reference intervals (RIs) present a reliable framework for interpreting hematology and clinical biochemistry results in avian and reptile patients. Specifically, RIs provide evidence-based support for veterinarians to make clinical decisions when comparing patients with the central 95% of reference sample populations; however, while RIs derived from a single, sizable healthy population serve a general interpretation framework, understanding how reference intervals may vary by major physiologic factors including sex, life stages, and seasonality enables more appropriate clinical decisions in captive exotic species. In this study, we examined these variations in hematology and clinical biochemistry RIs using wellness patient populations relegated to cockatoos and bearded dragons.

Introduction

The interpretation of quantitative laboratory results such as hematology and clinical biochemistry relies on establishing clinical reference intervals. Veterinarians can reasonably discriminate between healthy and abnormal states in novel subjects by comparing the patient's clinical values to a range of low and high reference limits derived from the central 95% of a known healthy population.[4]

The construction of RIs for an exotic species using the traditional direct approach follows three major steps. First, a reference population is selected using predefined wellness criteria, then measurements are conducted. Finally, reference intervals are derived using parametric and non-parametric statistical methods.[11] The selected population is often condition-free, but RIs can be established for any well-defined population. Ideally, a minimum of 120 samples are required to create robust clinical RIs. This stringent standard reduces the effects a small number of patients with sub-clinical diseases and outliers may have on RIs.[1,9]

There are unique challenges to provisioning RIs in exotic species. Predefining and selecting a sufficiently healthy population of any unique exotic species is difficult given the diversity in exotics.[4] Moreover, different study conditions such as the time of year or geographic locations may induce biases. In all, while RIs derived from a single population of healthy animals provide a general interpretation framework, different physiologic and lifestyle factors are known to mediate the baseline RIs for many species.[12,13,16,18,19]

This study applied an indirect non-parametric RI framework to examine the dynamics of hematology and clinical biochemistry RIs in two captive exotic populations: cockatoos (family Cacatuidae) and bearded dragons (species *Pogona vitticeps*). The wellness populations were stratified by sex and modeled smoothly through stages of life and time of year. We demonstrated that 1) the indirect RI derivation approach produced similar reference limits previously reported for cockatoos and bearded dragons 2) there may be significant differences in RIs depending on the factors used for population stratification and 3) while the difference in RIs through life stages and seasons can be discretized, the result shows that the change in RIs may be understood as continuous dynamics. Henceforth in this work, clinical variables and analytes were used interchangeably.

Materials and Methods

Sample collection

This study focused on RIs for cockatoos (family Cacatuidae) and bearded dragons (species *Pogona vitticeps*). We utilized an indirect RI approach to overcome the difficulty of sampling condition-free reference exotic populations. The main difference in the indirect RI derivation approach is that the reference intervals are determined a posteriori by statistical methods based on identifying a distribution in the midst of the data. This is contrary to the traditional approach which requires the assessment of individual results belonging to the pre-selected reference population.[11]

Routine wellness blood samples for cockatoos and bearded dragons from Moichor's partner clinics were drawn; hematology and clinical biochemistry results were processed and aggregated. Furthermore, Moichor hematology system and pathology findings by boarded pathologists confirmed the wellness status of the sample. Table 1 presents the distribution of samples by sex.

Table 1. Sample distribution by species and sex.

Animal	n (total)	n (male)	n (female)
Cockatoo	550	238	312
Bearded dragon	251	109	142

Hematology

For each collected sample, a blood smear was created and stained with modified Wright-Giemsa stain and processed using Moichor's Automated Hematology System.[14] The system operates by automatically classifying and quantifying leukocyte subtypes (WBC, Heterophils, Lymphocytes, Monocytes, Basophils, Eosinophils, Azurophils) under 400x magnification; subsequently, absolute leukocyte count and differential were produced and recorded.

Clinical chemistry

Plasma samples for clinical chemistry were collected in plasma separating tubes with light green tops. After the collection of whole blood samples at Moichor's partner clinics, samples were centrifuged to separate plasma from whole blood. A standard biochemistry profile was determined, including the concentration of total protein, albumin, AST, total bile acid, creatinine, uric acid, phosphorus, glucose, calcium, creatine kinase, cholesterol, and total triglycerides. The selected analytes were determined using colorimetric methods as measured by the Beckman Coulter AU680 Chemistry Analyzer (Beckman Coulter Inc, Brea, CA USA).

Statistical analyses

Statistical analyses were performed per individual hematology and biochemistry analytes. Because outliers artificially broaden the calculated reference intervals,[9] the outliers were removed from the aggregate data using the Interquartile Range Rule (IQR Method).[2] The remaining data for each analyte was aggregated by patient's sex; the patient's age was calculated based on the declared birth date, and the time of year or seasonality was determined by the number of days since the first day of the calendar years that a given sample was collected.

The reference interval for each analyte was calculated in aggregate based on how the clinical values are distributed.[3] For most clinical variables, an RI could be determined nonparametrically between 2.5th and 97.5th percentiles of the underlying distribution of values. On the other hand, when an analyte of interest or its transformation follows a Gaussian distribution, the parametric method can be applied, and the reference interval is to be within 1.96 SD around the distribution mean. Conceptually, both nonparametric and parametric methods of RI derivation attempt to determine the central 95% of clinical value distribution for any given analytes. While the nonparametric method is far simpler than estimating the RI based on the observed data and is generally preferred, the parametric method allowed for interrogating the dynamics of RIs in different analytes.

While most investigations into the seasonal or age dynamics in RI discretized the data into measurement groups for ease of experimentation and sufficient sample size.[12,13,16,18,19] it is likely that these changes take place in a more transient manner. Gaussian Process Regression (GPR),[17] a flexible Bayesian nonparametric model, could be used to model and visualize the smooth reference range dynamic for each analyte. For the purpose of examining the dynamics of RIs based on age and time of year, the clinical variables were observed and assumed to be approximately Gaussian within smaller epochs. GPR had several benefits: First, it worked well on limited data. Second, GPR gave the ability to provide uncertainty measures in terms of variance; in the context of RI, this corresponded to the definition of reference interval for Gaussian distributed variables. Third, when indexed by time, GPR is autoregressive, thus enabling forecasting or estimating of reference ranges over sparser sampling locations. In this study, GPR was fit to individual analyte datasets through stages of life and time of year. The kernel used in the regression model is a composite kernel[8]: the sum of Exponentiated Squared Kernel (RBF) and White Kernel.

Results

Static reference intervals are in agreement with previously reported RIs. For each of the two patient populations, static RIs were reported for males, females, and in aggregate. Furthermore, the distribution of each clinical value was also reported. In both captive populations, the static reference ranges were similar for most analytes (Fig 1). Furthermore, these RIs overlap with previously reported reference limits for these species.[7,15]

Static male and female RIs are in agreement. Without examining the RI dynamics, distributional characteristics for clinical values are similar between male and female patient groups for each clinical variable. (Figure 2).

Cockatoo	RI Male	RI Aggregate	RI Female
WBC (/uL)	4945 - 16383	4994 - 16810	5034 - 17327
Heterophil (/uL)	2514 - 11788	2475 - 11530	2445 - 11429
Lymphocyte (/uL)	798 - 5941	878 - 5902	905 - 5888
Monocyte (/uL)	0 - 788	0 - 759	0 - 755
Basophil (/uL)	0 - 538	0 - 535	0 - 517
Eosinophil (/uL)	0 - 389	0 - 422	0 - 437
Heterophil Diff (%)	0.38 - 0.85	0.39 - 0.85	0.39 - 0.84
Lymphocyte Diff (%)	0.11 - 0.54	0.11 - 0.55	0.11 - 0.55
Monocyte Diff (%)	0.00 - 0.09	0.00 - 0.09	0.00 - 0.09
Basophil Diff (%)	0.00 - 0.07	0.00 - 0.06	0.00 - 0.06
Eosinophil Diff (%)	0.00 - 0.04	0.00 - 0.05	0.00 - 0.05
Total Protein (g/dL)	3.0 - 4.2	2.9 - 4.1	2.9 - 4.1
Albumin (g/dL)	1.8 - 2.8	1.8 - 3.0	1.8 - 3.0
Asparatate Aminotransferase (U/L)	99 - 214	98 - 226	97 - 230
Total Bile Acid (umol/L)	9.2 - 68.4	10.3 - 76.5	10.6 - 77.8
Creatinine (mg/dL)	0.20 - 0.42	0.20 - 0.44	0.19 - 0.44
Uric Acid (mg/dL)	2.1 - 10.9	2.2 - 10.7	2.3 - 10.7
Phosphorus (mg/dL)	1.59 - 5.58	1.62 - 5.69	1.69 - 5.81
Glucose (mg/dL)	205 - 311	208 - 309	211 - 305
Calcium (mg/dL)	8.15 - 10.36	8.17 - 10.38	8.20 - 10.44
Creatine Kinase (U/L)	125 - 962	127 - 905	128 - 885
Cholesterol (mg/dL)	176 - 359	177 - 359	180 - 356
Total Triglycerides (mg/dL)	71 - 208	70 - 209	70 - 211

Bearded Dragon	RI Male	RI Aggregate	RI Female
WBC (/uL)	5395 - 27830	6149 - 29095	6353 - 29187
Heterophil (/uL)	1014 - 6388	979 - 9641	924 - 8051
Lymphocyte (/uL)	2300 - 18939	2182 - 18978	2184 - 18952
Monocyte (/uL)	0 - 577	0 - 582	0 - 594
Basophil (/uL)	182 - 1538	169 - 1565	170 - 1583
Eosinophil (/uL)	0 - 245	0 - 287	0 - 285
Azurophil (/uL)	91 - 1622	109 - 1769	132 - 1900
Heterophil Diff (%)	0.08 - 0.52	0.09 - 0.52	0.09 - 0.51
Lymphocyte Diff (%)	0.31 - 0.85	0.31 - 0.85	0.32 - 0.83
Monocyte Diff (%)	0.00 - 0.03	0.00 - 0.05	0.00 - 0.05
Basophil Diff (%)	0.01 - 0.13	0.01 - 0.13	0.01 - 0.13
Eosinophil Diff (%)	0.00 - 0.03	0.00 - 0.03	0.00 - 0.03
Azurophil Diff (%)	0.01 - 0.11	0.01 - 0.12	0.01 - 0.13
Total Protein (g/dL)	3.9 - 6.8	3.7 - 6.8	3.7 - 6.8
Albumin (g/dL)	2.8 - 5.6	2.7 - 5.6	2.7 - 5.7
Asparatate Aminotransferase (U/L)	13 - 64	13 - 80	13 - 88
Total Bile Acid (umol/L)	2.4 - 41.2	1.7 - 10.5	1.5 - 35.0
Uric Acid (mg/dL)	2.1 - 11.3	146.5 - 268.7	1.7 - 9.7
Phosphorus (mg/dL)	2.69 - 8.03	10.79 - 21.86	2.96 - 11.25
Glucose (mg/dL)	153 - 251	142 - 172	143 - 285
Calcium (mg/dL)	10.96 - 23.82	10.79 - 21.86	10.77 - 20.94
Creatine Kinase (U/L)	80 - 4354	70 - 4136	70 - 3262
Cholesterol (mg/dL)	156 - 456	158 - 451	163 - 443
Total Triglycerides (mg/dL)	60 - 520	51 - 527	49 - 521

Figure 1. Hematology and clinical biochemistry RIs calculated based on aggregated wellness cockatoo samples (left) and bearded dragon (right).

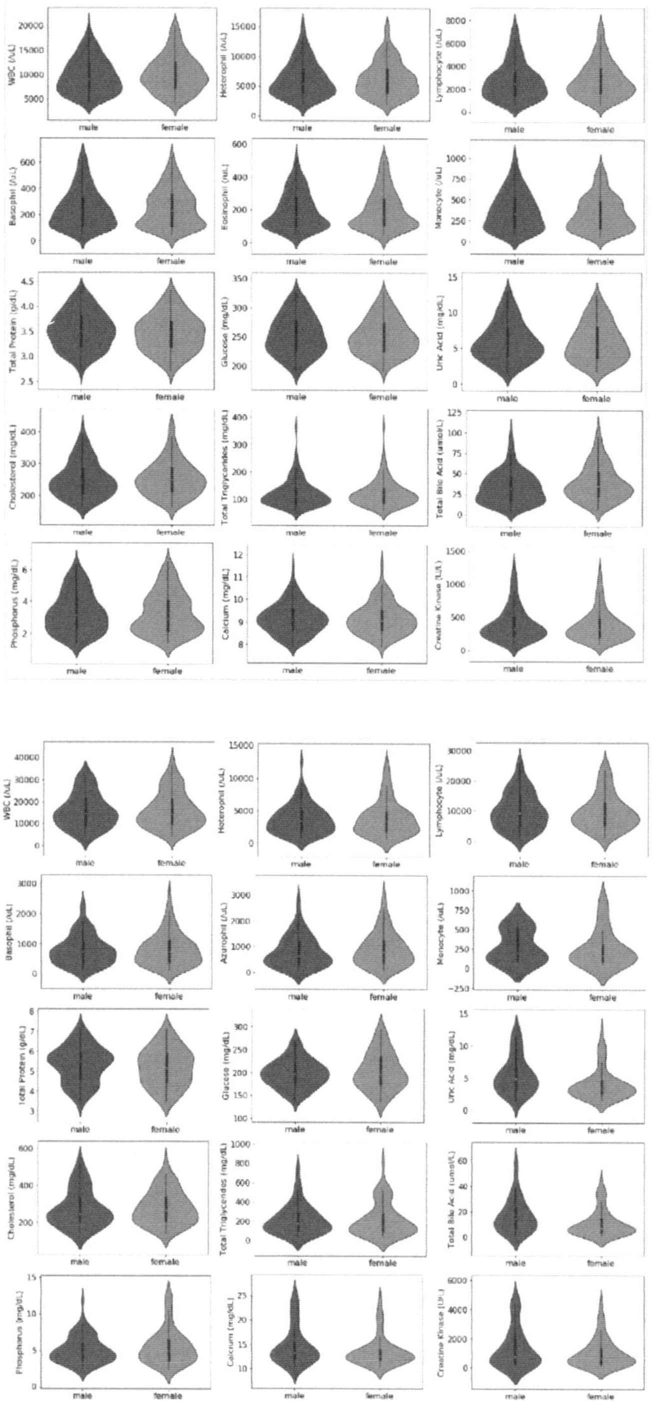

Figure 2. Distribution of hematology and clinical biochemistry values in male vs female wellness cockatoo samples (top panels) and bearded dragons (bottom panels). Each sub-figure represents a unique analyte.

Smooth dynamic RIs better reflected known seasonal trends and events. Like other parrots, captive cockatoos molt in the Spring and Fall during a process that may last 8-12 weeks.[5] During molting periods, higher WBC, heterophils, and lymphocyte counts were observed (Fig 3). This finding was consistent with previous studies on cockatoos and other avian species during their molting periods. This modulation in immune response was thought to be shaped by a trade-off between the availability of resources and the probability of encountering pathogens.[5,6] Moreover, the change in serum total protein level along with glucose and calcium (Fig 4) was also consistent with prior studies. Compared to the discretized reference intervals, the smooth dynamical RIs demonstrated the transient nature of cockatoo physiologic events. While we observed similar seasonal dynamics in bearded dragons (Figs 5 and 6), growing data would continue to highlight major seasonal trends.

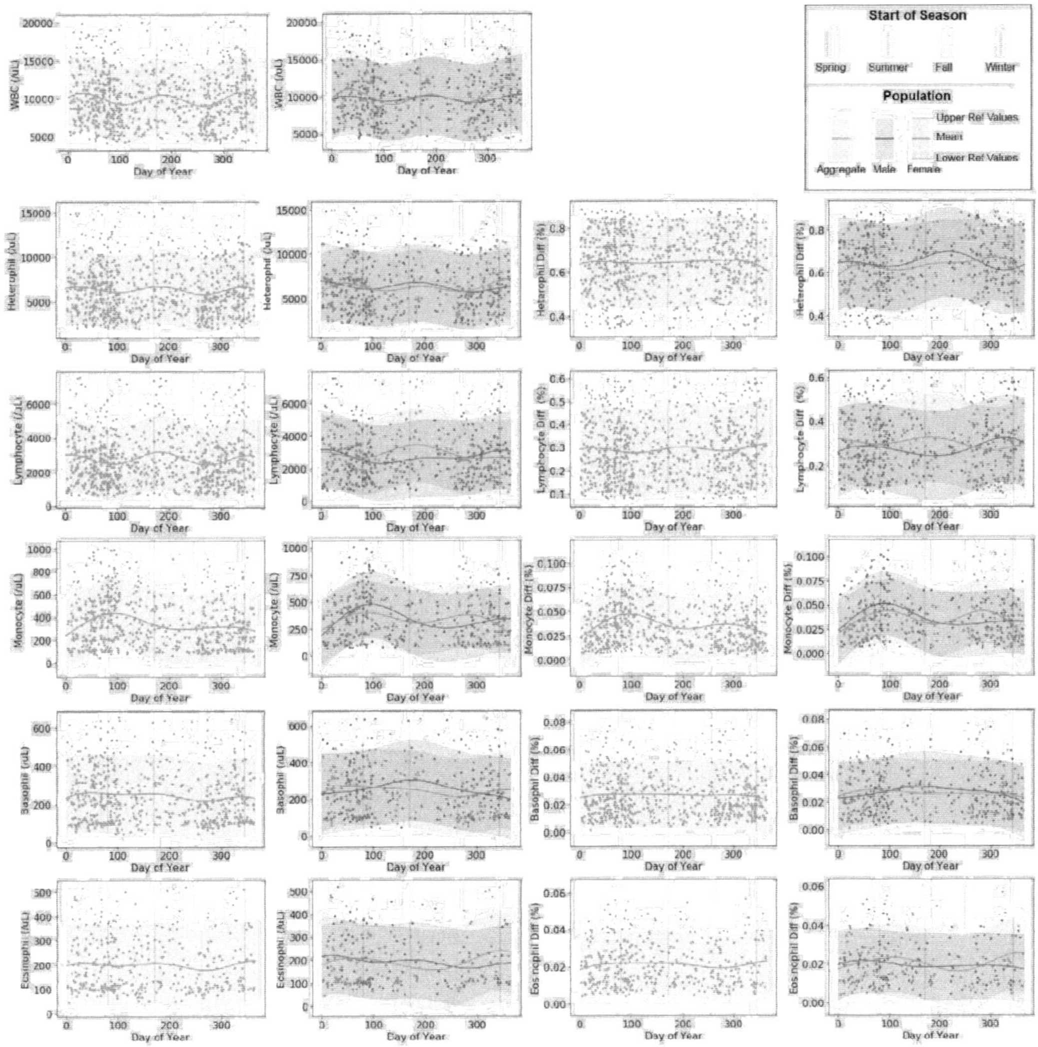

Figure 3. GPR Mean and 95% reference range of hematology variables in cockatoos through the seasons. Each variable consists of two panels representing regression on days since the beginning of the year using aggregate data (left) and data stratified by sex (right). Vertical lines in each sub-figure represent the beginning of each meteorologic season in the United States: Spring, Summer, Fall, and Winter.

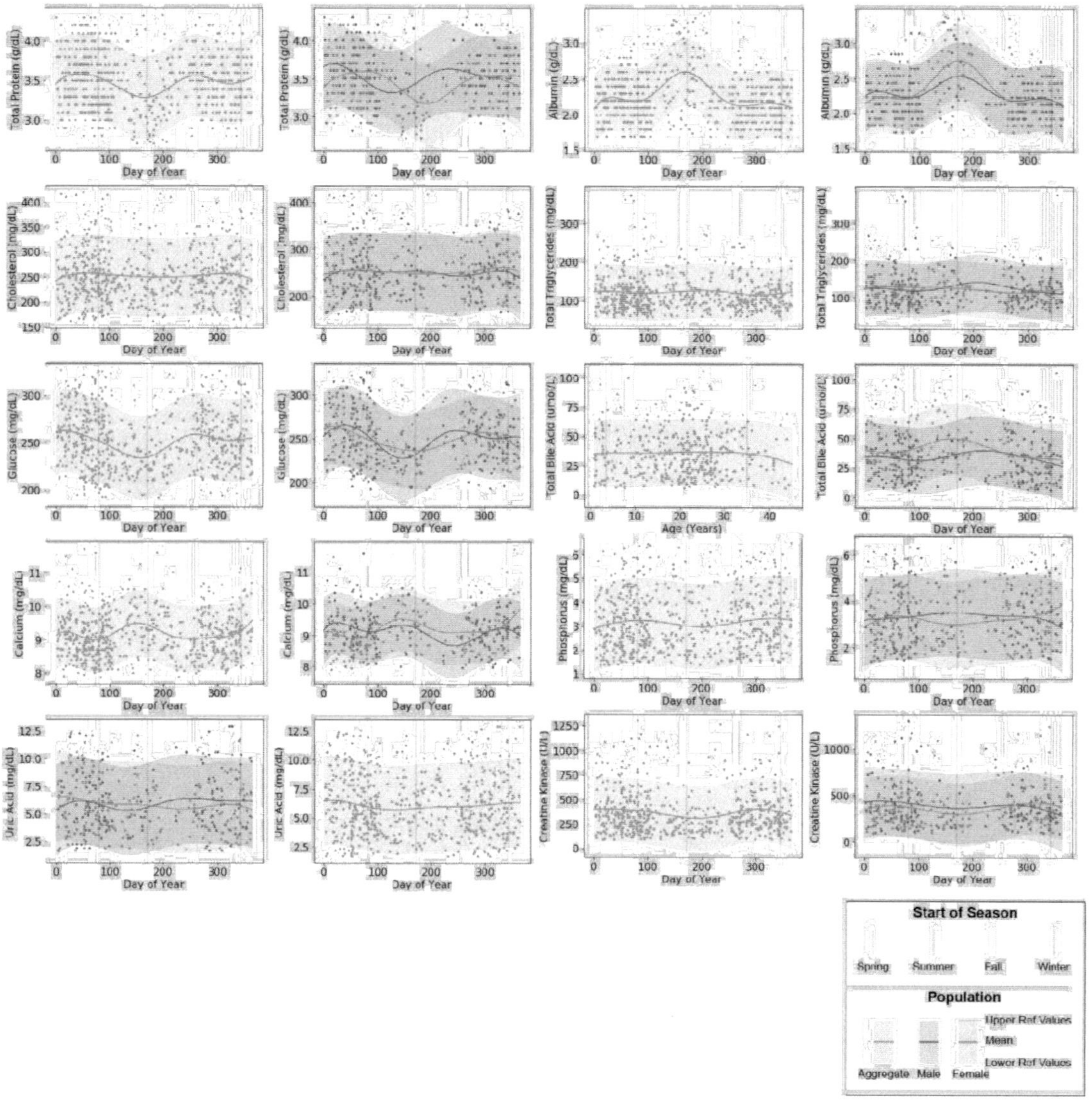

Figure 4. GPR Mean and 95% reference range of clinical biochemistry variables in cockatoos through seasons. Each variable consists of two panels representing regression on the day since the beginning of the year using aggregate data (left) and data stratified by sex (right). Vertical lines in each sub-figure represent the beginning of each meteorologic season in the United States: Spring, Summer, Fall, and Winter.

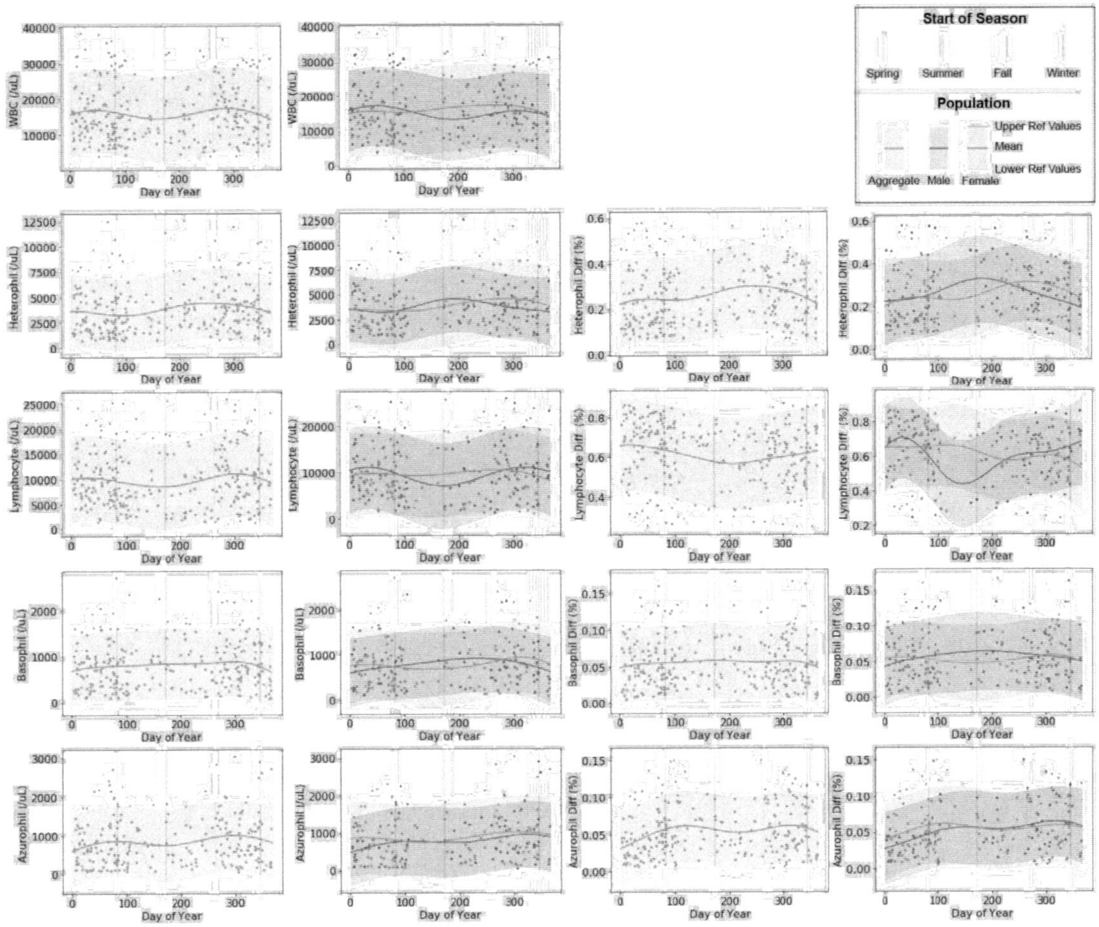

Figure 5. GPR Mean and 95% reference range of hematology variables in bearded dragons throughout the seasons. Each variable consists of two panels representing regression on the day since the beginning of the year using aggregate data (left) and data stratified by sex (right). Vertical lines in each sub-figure represent the beginning of each meteorologic season in the United States: Spring, Summer, Fall, and Winter.

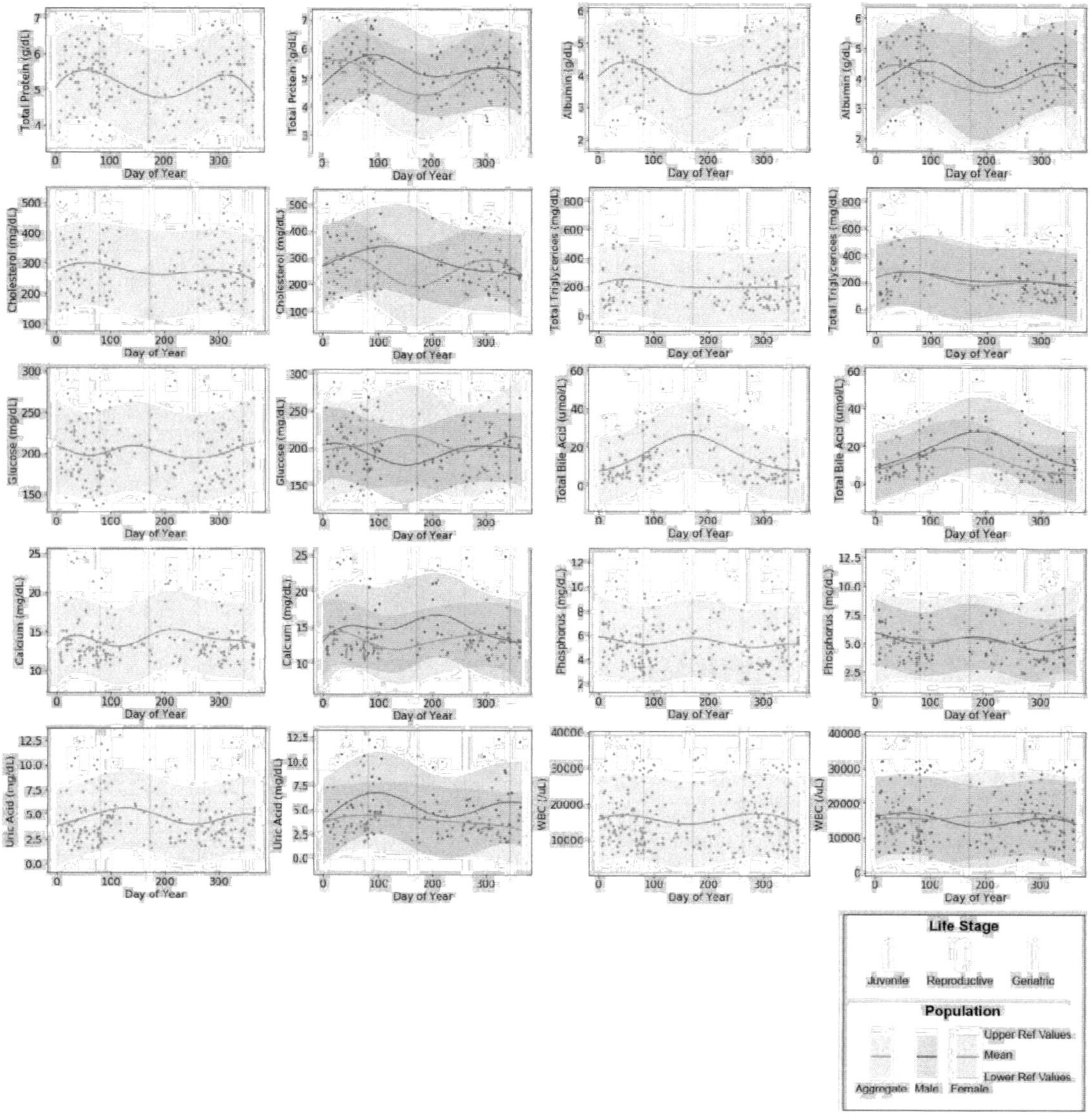

Figure 6. GPR Mean and 95% reference range of clinical biochemistry variables in bearded dragons throughout the seasons. Each variable consists of two panels representing regression on the day since the beginning of the year using aggregate data (left) and data stratified by sex (right). Vertical lines in each sub-figure represent the beginning of each meteorologic season in the United States: Spring, Summer, Fall, and Winter.

Smooth dynamic RIs showed marked changes in aging individuals. While the RI data suggest a relatively stable hematology and chemistry RIs for both adult cockatoos and bearded dragons, there was a more drastic shift in the baseline values for patients entering the geriatric stage of life in both reference populations (Figs 7-10).

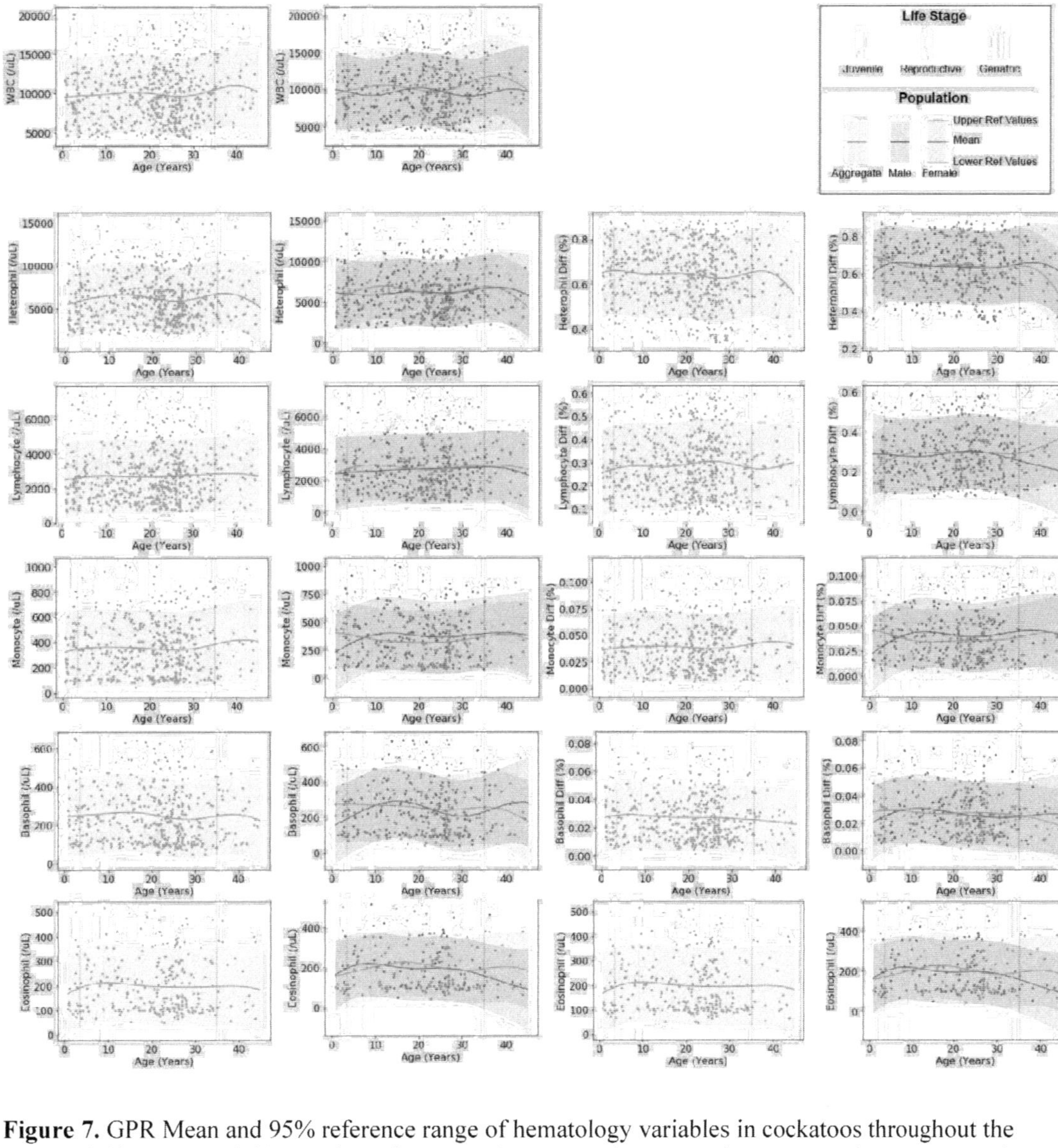

Figure 7. GPR Mean and 95% reference range of hematology variables in cockatoos throughout the stages of life. Each variable consists of two panels representing regression on age using aggregate data (left) and data stratified by sex (right). Vertical lines in each sub-figure represent the beginning of different stages of life: juvenile, reproductive, and geriatric.

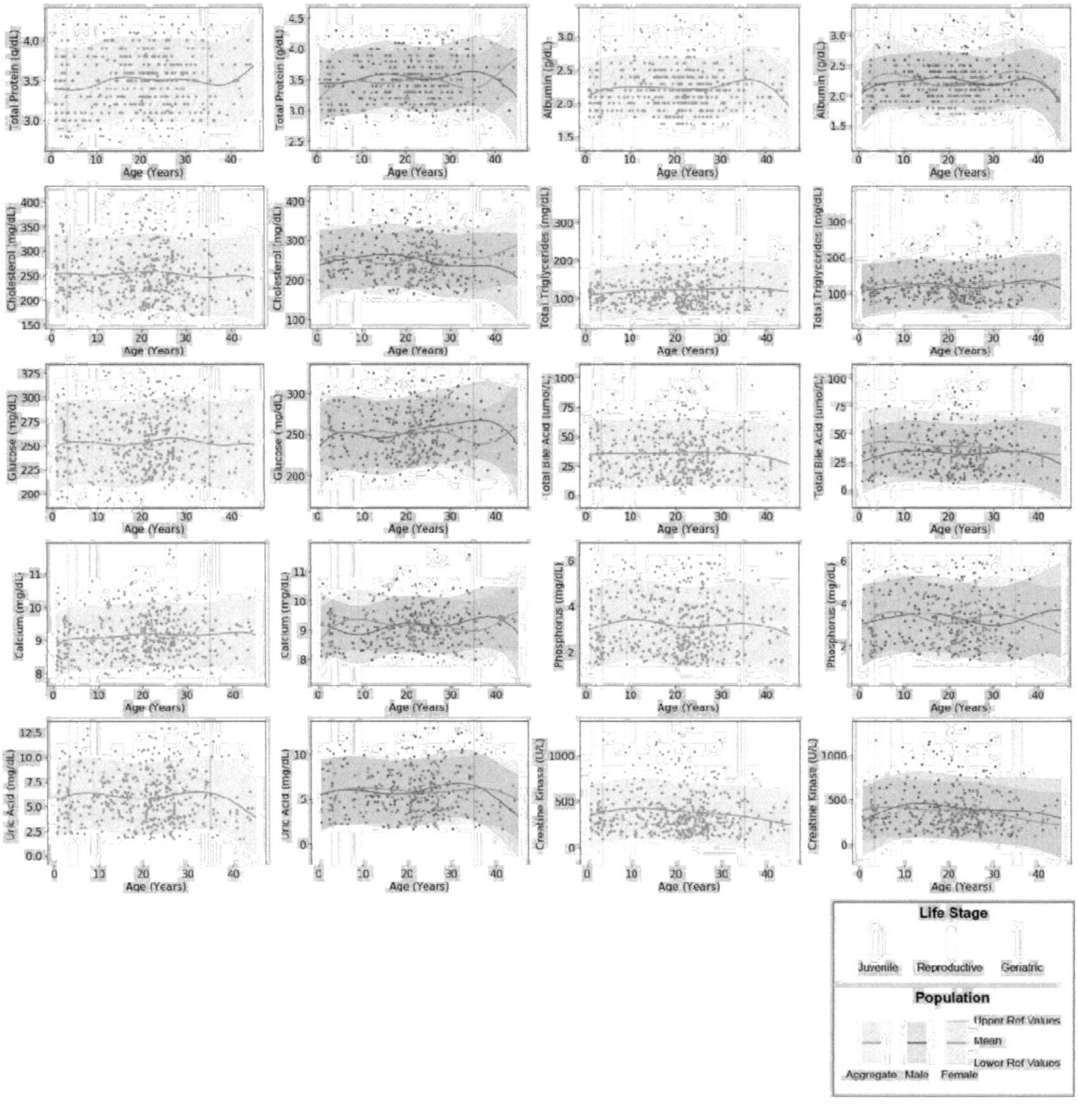

Figure 8. GPR Mean and 95% reference range of clinical biochemistry variables in cockatoos through the stages of life. Each variable consists of two panels representing regression on age using aggregate data (left) and data stratified by sex (right). Vertical lines in each sub-figure represent the beginning of different stages of life: juvenile, reproductive, and geriatric.

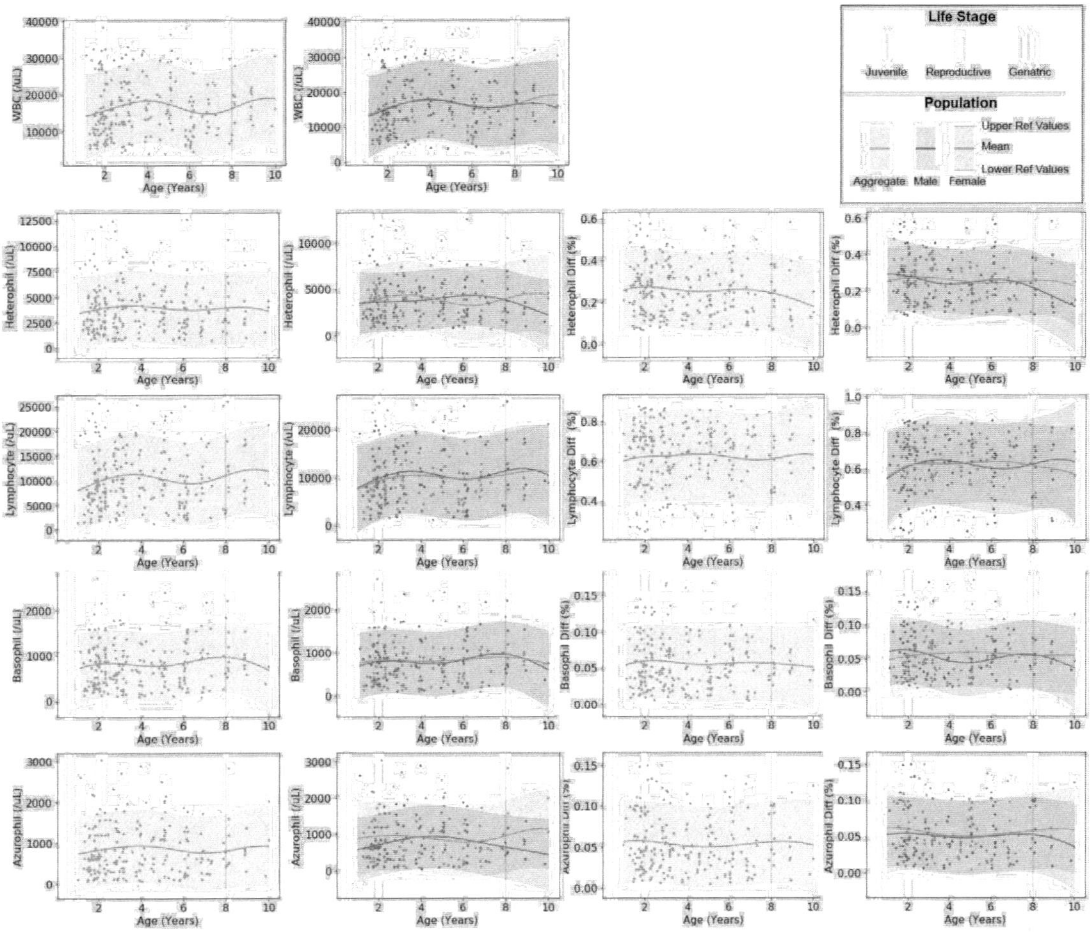

Figure 9. GPR Mean and 95% reference range of hematology variables in bearded dragons through stages of life. Each variable consists of two panels representing regression on age using aggregate data (left) and data stratified by sex (right). Vertical lines in each sub-figure represent the beginning of different stages of life: juvenile, reproductive, and geriatric.

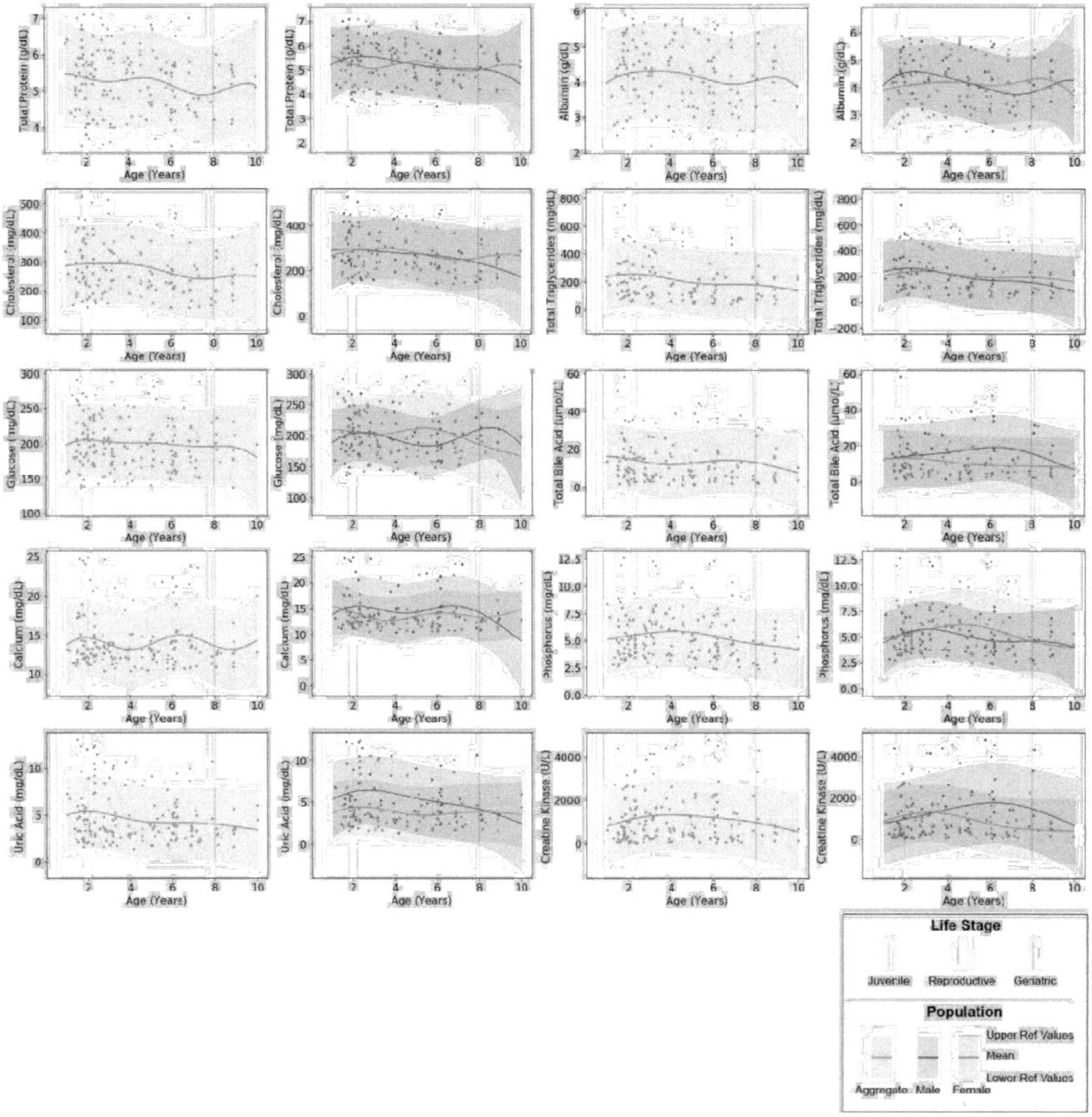

Figure 10. GPR Mean and 95% reference range of clinical biochemistry variables in bearded dragons through stages of life. Each variable consists of two panels representing regression on age using aggregate data (left) and data stratified by sex (right). Vertical lines in each sub-figure represent the beginning of different stages of life: juvenile, reproductive, and geriatric.

Discussion

In this study, we demonstrated the use of an indirect approach to construct dynamic reference intervals for cockatoos and bearded dragons. The resulting aggregate RIs significantly overlapped with the previously established reference limits using the traditional approach. Furthermore, we showed that while discretely constructed reference intervals are appropriate for many population stratifications such as by sex, seasonal, and life stages, RIs might be better conceptualized and understood as continuous dynamics.

The study populations were primarily companion animals; thus, these patients were kept in relatively controlled environments. Even though the population was captive, we could capture intrinsic seasonal and life stage changes through the construction of dynamic reference intervals. It would be informative to compare the differences in dynamic RIs between wild and captive/companion populations.

In addition, the study population originated from 41 states across the United States, resulting in a wide range of geographic sampling locations for the indirect reference populations. While we do not expect dramatic effects on reference intervals due to the geographic distribution of samples, we intend to explore these differences in RIs in the future.

It is crucial to note the assumption that clinical variables are approximately Gaussian in smaller time epochs. Although this assumption allowed us to efficiently model the dynamics of baseline values in most analytes for this study, in clinical practice, it would be essential to apply appropriate mathematical transformations to ensure the integrity of RIs built on this Gaussian assumption.[3,10]

Perhaps, the most important implication of this study is how the proposed dynamic RIs would affect veterinarians' understanding of patients' state of wellness. Future studies will focus on how each patient's intervention threshold would be affected due to differential diagnoses resulting from shifts in reference intervals.

References

1. American Society for Veterinary Clinical Pathology. Quality Assurance and Laboratory Standards Guidelines. Guidelines for the determination of reference intervals (RI) in veterinary species. 2022. Available at: https://cdn.ymaws.com/www.asvcp.org/resource/resmgr/QALS/Other_Publications/RI_Guidelines_For_ASVCP_webs.pdf. Accessed June 27, 2022.

2. Chandola V, Banerjee A, Kumar V. Outlier detection: a survey. ACM Computing Surveys, 14:15. 2007. Available at; https://www.researchgate.net/profile/Vipin-Kumar-54/publication/242403027_Outlier_Detection_A_Survey/links/0deec52d83600bb86c000000/Outlier-Detection-A-Survey.pdf. Accessed June 27, 2022.

3. Clinical and Laboratory Standards Institute. *Defining, Establishing, and Verifying Reference Intervals in the Clinical Laboratory; Approved Guideline.* 3rd Ed. Wayne, PA: CLSI; 2010.

4. Cray C. Reference intervals in avian and exotic hematology. *Vet Clin North Am Exot Anim Pract.* 2015;18(1):105-116.

5. Dawson A. Avian molting. In: Scanes CG, ed. *Sturkie's Avian Physiology.* St. Louis, MO: Elsevier; 2015: 907-917.

6. Demina I, Tsvey A. Total white blood cell counts but not HL ratio changes in the transition from post-juvenile molt to autumn migration in the first-year Eurasian blackcap (*Sylvia atricapilla*). *J Exp Zool A Ecol Integr Physiol.* 2022;337(2):159-170.

7. Doneley B, Monks D, Johnson R, Carmel B. Reference intervals for commonly kept reptile species. In: Doneley B, Monks D, Johnson R, Carmel B, eds. *Reptile Medicine and Surgery in Clinical Practice.* Ames, IA: Wiley-Blackwell; 2018:473-479.

8. Duvenaud D, Lloyd JR, Grosse R, et al. Structure discovery in nonparametric regression through compositional kernel search. *Proc Int Conf Machine Learn.* 2013;28(3):2203-2211.

9. Horn PS, Feng L, Li Y, Pesce AJ. Effect of outliers and nonhealthy individuals on reference interval estimation. *Clin Chem.* 2001;47(12):2137-2145.

10. Horn PS, Pesce AJ, Copeland BE. A robust approach to reference interval estimation and evaluation. *Clin Chem.* 1998;44(3):622-631.

11. Jones GR, Haeckel R, Loh TP, et al. Indirect methods for reference interval determination-review and recommendations. *Clin Chem Lab Med.* 2018;57(1):20-29.

12. Kasprzak M, Hetmanski T, Kulczykowska E. Changes in hematological parameters in free-living pigeons (*Columba livia* f. *urbana*) during the breeding cycle. *J Ornithol.* 2006;147(4):599-604.

13. LaGrange SM, Kimble SJ, MacGowan BJ, Williams RN. Seasonal variation in hematology and blood plasma chemistry values of the timber rattlesnake (*Crotalus horridus*). *J Wildl Dis.* 2014;50(4):990-993.

14. Le T, Lazo A, Powers L. Automation of avian and reptile complete blood count using artificial intelligence powered microscopy. *Proc ExoticsCon.* 2021;517-518.

15. Ritchie BW, Harrison GJ, Harrison LR, eds. *Avian Medicine: Principles and Application.* Lake Worth, FL: Wingers Publishing;, 1994.

16. Pages T, Peinado VI, Viscor G. Seasonal changes in hematology and blood chemistry of the freshwater turtle *Mauremys caspica leprosa. Comp Biochem Physiol A.* 1992;103(2):275-278.

17. Rasmussen CE, Williams CKI. Gaussian Process for Machine Learning. 2006. Available at: https://gaussianprocess.org/gpml/. Accessed June 27, 2022.

18. Schmidt EMDS, Paulillo AC, Santin E, et al. Hematological and serum chemistry values for the ring-necked pheasant (*Phasianus colchicus*): variation with sex and age. *Int J Poult Sci.* 2007;6(2):137-139.

19. Silva LFN, Riani-Costa CCM, Ramos PRR, Takahira RK. Seasonal influence on biochemical profile and serum protein electrophoresis for boa constrictor amarali in captivity. *Braz J Biol.* 2011;71(2):517-520.

Comparison of Ear Microbiome in Rabbits with and without Otitis Externa using Next-Generation DNA Sequencing

Gina Vecere, DVM,
Shahar Malka, DVM, DABVP (Avian Practice),
Natasha Holden, MSc,
Shuiquan Tang, PhD,
Janina A. Krumbeck, PhD

Session #2507

Affiliation: From the Long Island Bird and Exotics Veterinary Clinic, 333 Great Neck Road, Great Neck, NY 11021, USA (Vecere, Malka), MiDOG LLC, 17062 Murphy Avenue, Irvine, CA 92614, USA (Holden, Tang, Krumbeck).

Note: Abstract format selected by author(s). See Front Matter for description of conference proceedings manuscripts submitted as scientific abstracts.

Abstract: Next-generation DNA sequencing (NGS) technologies have been increasingly explored in veterinary medicine in recent years, providing new and practical opportunities for clinical case intervention.[1,2] In this study we aimed to characterize the normal external ear canal microbiome of clinically healthy rabbits compared to that of rabbits presenting with otitis externa. Swabs from the external ear canal of 34 clinically healthy rabbits and 16 rabbits diagnosed with otitis externa were collected. Next-generation DNA sequencing was used to analyze samples for their respective bacterial and fungal microbiomes. A variety of previously unrecognized organisms were noted in analyses of both the clinically healthy and otitis externa groups. These organisms did not align with those traditionally considered to be the primary organisms in the microbiome of the external ear canal of rabbits, for clinically healthy or disease state ears.[3-5] In addition, there was a significantly higher number of different bacterial and fungal species in the clinically healthy group compared to the otitis group ($P = 0.002$ and $P < 0.001$, respectively). This study provided evidence that the rabbit aural microbiome profile is distinctly different between a clinically healthy and an otitis state. It also highlighted new bacterial and fungal organisms of note in rabbits diagnosed with otitis externa compared to those previously thought to be the primary disease-causing organisms. Understanding the microbial population dynamics in the rabbit aural microbiome is particularly critical for helping clinicians recognize, prevent, or revert the progression of otitis.

References

1. Tang S, Prem A, Tjokrosurjo J, et al. The canine skin and ear microbiome: A comprehensive survey of pathogens implicated in canine skin and ear infections using a novel next-generation-sequencing-based assay. *Vet Microbiol*. 2020;247:108764.

2. Krumbeck J, Reiter A, Pohl J, et al. Characterization of oral microbiota in cats: novel insights on the potential role of fungi in feline chronic gingivostomatitis. *Pathogens*. 2021;10(7):904.

3. Mancinelli E, Lennox A. Management of otitis in rabbits. *J Exot Pet Med*. 2017;26(1):63-73.

4. Oglesbee BL. Otitis externa and media. In: Oglesbee BL, ed. *The 5-Minute Veterinary Consult: Ferret and Rabbit*. Ames, IA: Wiley-Blackwell; 2006:308-311.

5. Quesenberry KE, Carpenter JW. Neurological and musculoskeletal diseases. In: Quesenberry KE, Carpenter JW, eds. *Ferrets, Rabbits, and Rodents: Clinical Medicine and Surgery.* 3rd Ed. St. Louis, MO: Elsevier Saunders; 2012:245-255.

Pododermatitis in Small Mammals and Birds

Nico Schoemaker, DVM, PhD, DipECZM (Small Mammal), DipECZM (Avian), and
Yvonne van Zeeland, DVM, MVR, PhD, DipECZM (Avian), DipECZM (Small Mammal), CPBC

Session #2657

Affiliation: From the Division of Zoological Medicine, Department of Clinical Sciences, Faculty of Veterinary Medicine, Utrecht University, Utrecht, 3584 CM, The Netherlands.

Previously published in Schoemaker N, van Zeeland YRA. Pododermatitis in birds and small mammals. Proc 8th World Congr Vet Dermatol Contin Educ. 2016;403-409.

Abstract: Pododermatitis is a common condition in small mammals and birds, characterized by lesions in areas where soft tissue structures overly the bony structures of the feet. Upon cross-comparison, many striking similarities can be identified. For example, lack of exercise, inadequate nutrition, overweight and improper perching or floor substrate are considered portent contributing factors in most species. As such, addressing the underlying cause is an important cornerstone of treatment, in addition to treating pain, secondary infections and managing the wound. Staging of the lesions is deemed helpful to aid in adequate decision-making on treatment and prognosis. While many cases can be managed with the proper care, wound healing may take a considerable time, with advanced cases often responding poorly to therapy. Educating clients on prevention of pododermatitis is therefore considered key to minimize the incidence of this condition in practice.

Introduction

Pododermatitis, also referred to as bumblefoot (in birds) or sore hocks (in small mammals), is a condition characterized by the presence of infection, inflammation and/or degenerative changes of the ventral surface of the feet. The condition is commonly seen in both small mammals and birds. Without appropriate therapy, the disease can quickly progress to not only involve the skin and connective tissues of the foot, but also the deeper tissues (tendons, joints and skeletal structures), thereby resulting in significantly decreased or loss of limb function.

Etiology and Pathogenesis

In both birds and small mammals, pododermatitis has been linked to continuous or uneven pressure applied to the plantar surface of the feet. This pressure may subsequently result in disruption of the epithelial surfaces and ischemia and pressure necrosis of the soft tissue structures overlying the bony structures of the feet. As a sequela to the ischemia and necrosis, inflammatory mediators are released, which, together with the oxidative damage caused during the intermittent reperfusion of the area, result in further tissue damage and vascular thrombosis. At this stage, chronic inflammation, hyperkeratosis, erosions and ulcerations may be seen. Invasion of bacterial organisms, in particular *Staphylococcus aureus* (all species), *Escherichia coli* (in raptors, parrots) and *Pasteurella multocida* (in rabbits), may subsequently result in secondary bacterial infections of the skin and underlying tissues. In severe cases, infection may spread to the underlying bones, tendons and joints, resulting in osteomyelitis, tenosynovitis and arthritis, thereby complicating the disease process and reducing the chances of the limb function returning to normal, especially if permanent damage to the tendons (e.g., displacement, rupture) is present. Occasionally, ulcerative lesions may also disrupt the integrity of the blood vessels of the foot, thereby resulting in significant hemorrhage and anemia. In advanced cases, pododermatitis may furthermore lead to systemic spread of the infection with subsequent endocarditis or polyarthritis.

Predisposing Factors

Many factors may play a role in the development of pododermatitis. In general, any management-related, medical or behavioral condition that results in an increased pressure of the weight bearing surfaces, alterations in the weight distribution or stance of the animal and/or increased friction and irritation of the skin may predispose an animal to developing pododermatitis (Table 1). Management-related factors that have been found to increase the risk for pododermatitis include lack of exercise, confinement to a small space, improper substrate or perches, poor or unsanitary housing and malnutrition. In addition, excess body weight, trauma and other physical or behavioral conditions that lead to an altered stance, activity and/or weight bearing (e.g., spondylosis, arthritis, neurologic disease, nervousness, conformational deformities) may increase the risk of the animal developing pododermatitis.

Table 1. Predisposing risk factors for pododermatitis in birds and small mammals.

Risk factors
• Obesity or, in rabbits, emaciation
• Pregnancy (in small mammals)
• Inadequate nutrition (e.g., hypovitaminosis A [psittacine birds], biotin deficiency [turkeys], hypovitaminosis C [guinea pigs])
• Lack of exercise
• Improper perches or substrate (e.g., inappropriately sized perches; abrasive surfaces such as carpets, sandpaper; hard floors such as tiles or cement; wire bottom cages)
• Moist and unhygienic conditions, leading to urinary and/or faecal soiling of the feet
• Poor husbandry (e.g., small enclosures, overcrowding)
• Unilateral lameness or pain, causing increased weight bearing to the contralateral limb (e.g., arthritis, fractures)
• Conditions that lead to reduced mobility, paresis, paralysis, or altered gait (e.g., spondylosis, neurologic disease, limb deformities)
• Trauma and abrasions to the feet (e.g., overgrown talons [raptors], puncture/bite wounds, trap injury, thermal and electrical burns/wounds, frostbite)
• Infections (e.g., *Staphylococcus aureus*; *Pasteurella multocida* [rabbits], poxvirus [raptors, poultry], *Knemodikoptes* [budgerigars]) – may both be the primary cause as well as secondary to erosion/ulceration of the plantar surface, thereby complicating disease
• Loss of protective hair covering (e.g., iatrogenic clipping of hair) in rabbits
• Behavioural: nervousness, excessive thumping (rabbits); bating of jesses (raptors)

Although pododermatitis may develop in any bird or small mammal, certain species or breeds have been found to be more susceptible. For example, large and giant rabbit breeds (weighing >5 kg), raptors (e.g., falcons, eagles, osprey, red-tailed hawks), penguins, larger ducks and swans, and heavy-bodied psittacine birds (e.g., Amazon parrots, budgerigars, cockatiels) appear to be predisposed. Moreover, pododermatitis appears to occur particularly in rabbit breeds that have fine, sparse hair on the metatarsus (e.g., the Rex), thereby offering little protection against local injury, ischemia and pressure necrosis. Similarly, guinea pigs and rats are susceptible to sustaining injury to the plantar and palmar surfaces of their feet because of their relatively thin skin (similar to rabbits) and lack of hair on their feet and metatarsi. A study concerning risk factors of pododermatitis in rabbits from the UK revealed age and gender as predisposing factors, whereby they found pododermatitis to be more common in female, older rabbits. Moreover, neutering appeared to predispose the animals to pododermatitis; however, it cannot be excluded that this effect is the result from bias resulting from these animals also being more likely to be overweight. In other small mammals and birds, no age or sex predilection has been identified.

Clinical Presentation and Staging

Pododermatitis encompasses a range of clinical presentations varying from mild erythema, swelling and/or superficial ulcerations to deep ulcerations with concurrent osteomyelitis, arthritis and/or tenosynovitis with loss of limb function. In birds, the plantar metatarsal pad or plantar digital pads are most commonly affected, whereas in rabbits and rodents the lesions are usually located on the caudal aspect of the tarsus and metatarsus and, less frequently, the metacarpus and phalangeal regions of the front limbs. In the initial stages, lesions in small mammals may be limited to focal hair loss (in rabbits), erythema, scaling and thinning of the skin of the plantar surface over the bony prominence of the metatarsus. In addition, erosion, dry crusty scabs or callous-like swellings, combined with mild to moderate inflammation may be noted in the earlier stages of the disease. In more advanced cases, inflammation may be more severe and also involve the deeper structures, with obvious ulceration and tissue necrosis. Secondary infections may occur as well and result in suppurative dermatitis, abscessation, tenosynovitis, osteomyelitis and/or arthritis. At this stage, loss of pedal function may also be observed, (e.g., due to rupture or displacement of a tendon). Dependent on the severity of the podermatitis, mild to non-weight-bearing lameness of the affected limb may be seen. If pododermatitis is present bilaterally, animals may also show reluctance to move.

In both raptors and rabbits, classification schemes have been designed which help to grade pododermatitis depending on the extent, severity, and prognosis (Tables 2 and 3).

Table 2. Grading system for pododermatitis in raptors.[1]

Grade	Description
Grade I	Early lesion of the plantar surface involving integument only, with no secondary infection of deeper tissues. A loss of papilla, erythema and/or hyperkeratosis may be seen. Prognosis is excellent.
Grade II	Infection of the subcutaneous tissues with no gross swelling of affected feet. Localized signs of inflammation/infection and/or ischemic necrosis. Prognosis is good.
Grade III	Infected, swollen, painful feet without apparent damage to deep vital structures. Serous or caseous fluid draining from fibrotic lesion. Prognosis is good to guarded.
Grade IV	Infection of deep vital structures such as tendons and bones, though pedal function is retained. Patients may have concurrent tenosynovitis, arthritis, and/or osteomyelitis. Prognosis is guarded to poor.
Grade V	End-stage disease with osteomyelitis, crippling deformity, and loss of pedal function. Prognosis is grave and euthanasia is often performed.

Table 3. Grading system for pododermatitis in rabbits.[2]

Grade	Description
Grade 0	No lesions.
Grade 1	A small, circular area on the plantar aspect of the metatarsal bone-calcaneus (mono or bilateral lesions), with minimal alopecia, minimal epidermal hyperaemia and/or hyperkeratosis of the skin, but with no evidence of infection or bleeding of underlying tissues.
Grade 2	Circumscribed area of varying size localised at the caudal plantar aspect of the metatarsal-calcaneal area or extending linearly along the plantar aspect of the cranial metatarsal area with alopecia, erythema and scaling of surrounding tissues.
Grade 3	Area of varying size focally ulcerated and with varying degree of keratinisation abnormalities. Infection of subcutaneous tissue present.
Grade 4	Full-thickness skin loss with swelling and necrotic debris may be present with infection of underlying tissues. Purulent exudates may be adherent to the lesions.
Grade 5	Severe infections with involvement of deep structures including bones and tendons with tenosynovitis, osteomyelitis and arthritis.
Grade 6	End-stage disease with loss of pedal function.

Diagnostic Approach

A definite diagnosis of pododermatitis can usually be made following the history and physical examination. Moreover, the history may reveal information on the potential risk factors such as nutrition, housing, bedding, and exercise. Similarly, a full and systematic physical examination is needed to discover concurrent diseases or predisposing factors for the pododermatitis. For this purpose, the patient should initially be observed in its own cage or enclosure, whereby any changes in gait or abnormalities in resting or perching posture are recorded. A body weight and body condition score should also be obtained. Upon physical examination, special attention should be paid to the plantar (and palmar) surfaces of the feet to evaluate the presence of alopecia (in rabbits), erythema, erosions, ulcerations, necrosis, swelling and/or exudate. In addition, an orthopedic and/or neurologic examination may be performed to assess the limb function and presence of (other conditions resulting in) pain and reluctance to move.

Diagnostic work-up may further include a complete blood cell count, serum chemistry panel, and urinalysis (if applicable) to identify concurrent diseases or predisposing factors. Moreover, impression smears and a culture and sensitivity of the affected area may help to determine involvement of bacteria and guide the choice of antibiotic therapy. Preferably, cultures are performed from swabs are taken from deep within the lesion or pieces of infected tissue. If the lesion has extended into the deeper layers, imaging is recommended to identify involvement of the underlying bone or joints. Radiographs will usually suffice for this purpose (and may also help to assess presence of underlying limb and/or spinal disorders). However, Computed Tomography or Magnetic Resonance Imaging may be considered to detect more subtle abnormalities of the bone and soft tissues.

Management and Prognosis

Management of pododermatitis often requires a multimodal approach, which, dependent on the severity of the lesions, may include a combination of management, medical and/or surgical interventions. In any patient, treatment should be aimed at correction of the underlying causes and concurrent disease conditions. In addition appropriate wound care and bandaging, topical medication, systemic antibiotics, analgesics and/or surgical intervention may be indicated.

Correction of underlying causes

Any underlying causes should be addressed appropriately. This generally involves adjustments to the animal's housing, flooring, bedding and/or nutrition. For small mammals, cage floors should be adequately padded and/or bedded to reduce pressure on the feet (e.g., using foam rubber, thick towels, hay or fiber bedding). Bedding should be cleaned regularly to prevent risk of contamination and subsequent infection of the pedal lesions. In perching birds, perches of various sizes and types (e.g., wood, rope, rubber) should be provided. In addition, perches may be wrapped in safe, soft materials (e.g., cotton padding or bandaging tape) to provide extra protection. For raptors, flat perches covered with artificial turf (for falcons) or round/elliptical perches made of hemp or sisal rope (for eagles, hawks, owls) may be used. As for small mammals, surfaces should be kept clean and dry, and changed frequently. In waterfowl, access to swimming water (preferably a pool with gradual slopes) is important. Moreover, soft flooring or mats should be provided to reduce the pressure on the animal's feet. Enclosures should furthermore be designed in such a way that movement, exercise and activity are stimulated. Enrichment (e.g., feeding stations, toys, foraging enrichment, tunnels, solid wheels) may also help to achieve this goal.

Aside from housing, correction of the diet is also important, especially in overweight animals. Moreover, dietary corrections may be aimed at resolving any potential nutritional deficiencies (e.g. hypovitaminosis C in guinea pigs, hypovitaminosis A in parrots) that may contribute to the problem.

Wound care

In general, pododermatitis lesions require open wound management and healing by secondary intent. It is important to keep the ventral foot surface clean and dry, especially if ulcerations are present. For small mammals, hair surrounding the lesions may be trimmed (though not down to the skin!) to prevent matting and contamination. In addition, the wound may be thoroughly cleaned using chlorhexidine or povidone iodine solutions (in the initial phases of treatment), or by soaking it in warm saline or lactated Ringer solution (in the later stages of treatment). Following wound irrigation, necrotic tissue may be removed.

Bandaging

Bandaging is usually necessary, except in mild cases. Their function is to prevent tissue desiccation and protect the wound against further pressure, trauma and contamination. Generally, bandages are composed of three layers: 1) a primary layer consisting of a dressing that is in direct contact with the lesion and provides protection as well as encourages drainage of exudate, delivery of medication and debridement of necrotic tissue; 2) a secondary layer that is composed of an absorptive material such as cast padding or cotton padding; and 3) a tertiary layer that keeps the bandage in place (e.g., Vetrap or similar product). In severe cases, additional protection may be applied in the form of interdigitating bandages, dental acrylic or polypropylene foam shoes, styrene plastic polymer foot casts, ball bandages and/or doughnut-shaped bandages. These bandages may all help to relieve pressure on prominent weight-bearing surfaces and redistribute pressure onto other surfaces. The type of bandage chosen usually depends on the stage of healing, For example, wet-to-dry bandages are often used in the initial drainage stage, whereas ball bandages, polypropylene shoes and interdigitating bandages may be used in the granulation stage, early and later stages of healing, respectively. Regardless of the type of dressing and bandage chosen, bandages should be changed regularly, starting with daily changes until the condition improves, following which the time between bandage changes can be prolonged to once every 2-3 days up to a week until the condition resolves. Preferably, both feet are bandaged to avoid development of pododermatitis of the other foot as a result of the shift in weight-bearing.

Topical medication

Topical medication may be used alone or in conjunction with a bandage. Examples of medications that can be used topically to treat pododermatitis include silver sulfadiazine, fusidic acid, mupirocin calcium, Manuka honey, antibiotic ointments (e.g., piperacillin, rifampicin, lincomycin), dimethyl sulfoxide (DMSO) and/or products containing proteolytic enzymes. Care must be taken when using topical antibiotics, because ingestion or absorption can result in toxicity in many species, whereas oil-based products may affect feather quality and impair thermoregulation in birds. Topical corticosteroids are generally not recommended because of their immunosuppressive effects and delayed wound healing. Footpad strengthening products may be applied during advanced stages of healing.

Systemic antibiotics

Systemic antibiotics may be indicated in case of persistent or deep infections (grade 2 dermatitis or higher). Selection is preferably based on a culture and sensitivity, but should this not be available, a broad-spectrum antibiotic with action against gram-positive and gram-negative bacteria (e.g., trimethoprim-sulfamethoxazole) may be chosen. Antibiotic-impregnated beads (containing e.g., gentamycin, tobramycin or amikacin) also offer an effective method of antibiotic delivery to an infected ischemic area, especially in birds that underwent aggressive surgical debridement

Analgesia

Analgesics comprise an important part of the treatment regimen. Generally, analgesia is provided through administration of nonsteroidal anti-inflammatory drugs (e.g., meloxicam, carprofen) and/or opioids (e.g., buprenorphine, butorphanol).

Surgical intervention

More advanced cases of pododermatitis (grade 3 or higher) often require aggressive surgical debridement to remove infected or necrotic tissue, promote drainage (especially in case of an abscess), stimulate formation of granulation tissue and improve circulation to the area. The wound may then be flushed with warm saline after which antibiotic impregnated beads may be left behind in the wound. In addition, bandages mag be placed to protect the wound and reduce pressure on the weight-bearing surface. Generally, wounds are left open to heal by secondary intent, but in some patients (primary) closure of the wound may be attempted after a granulation bed has developed. In severe cases (e.g., pododermatitis in conjunction with tendon displacement, severe osteomyelitis, septic arthritis and/or loss of limb function), limb amputation may be considered as a salvage procedure (Note: not advisable for wild birds!).

In addition to surgical intervention, low-level laser therapy or therapeutic laser has also been found to improve wound healing by promoting fibroblast development, collagen production, and epithelialization; enhancing leukocyte infiltration and macrophage activity; promoting angiogenesis and stimulating vasodilatation; and increasing lymphatic drainage of the region.

Follow-Up and Prognosis

Pododermatitis can be challenging and difficult to treat. Particularly chronic pododermatitis will take several months to heal, depending on the severity and response to treatment. Prior to initiating therapy, owners should always be properly informed about the probable course and outcome. For mild to moderate cases of pododermatitis (grade 1-3) prognosis for recovery is generally considered good, especially if the underlying causes can be corrected. For grade 4-5 pododermatitis, the prognosis is considered guarded to poor, especially if osteomyelitis, septic arthritis and/or tenosynovitis are present. Euthanasia may be warranted in animals that suffer from severe, chronic pododermatitis, especially if more than one limb is affected, concurrent disease is present and/or the animal responds poorly to therapy and/or is not able to stand without severe pain.

Throughout the course of treatment, regular monitoring is advised whereby daily to weekly or biweekly rechecks are advised, dependent on the severity and stage of the disease process. Measuring the size of the lesion and documentation of the progress with the help of photographs are useful to evaluate the success or failure of the treatment plan. Aside from evaluating the lesion, assessment of the patient's general condition, as well as the efficacy of the used protocol for pain relief, is considered essential.

Prevention

Proper husbandry, diet and client education are important tools to help prevent the development of pododermatitis. This includes the implementation of correct husbandry practices (e.g., using various perch types and materials, suitable flooring and soft, dry bedding, sufficient space), proper sanitation, offering the animal environmental enrichment and exercise (including flight in birds of prey) as well as a nutritiously balanced diet for the species in question, and preventing the animal from becoming overweight. In addition, periodic evaluation of the animal's feet may help to detect pedal lesions in an early stage, thereby increasing the chances of successful intervention.

Selected References

1. Remple JD. Raptor bumblefoot: a new treatment technique. In: Redig PT, Cooper JE, Remple JD, et al, eds. *Raptor Biomedicine.* Minneapolis, MN: University of Minnesota Press; 1993:154-160.

2. Mancinelli E, Keeble E, Richardson J, Hedley J. Husbandry risk factors associated with hock pododermatitis in UK pet rabbits (*Oryctolagus cuniculus*). *Vet Rec.* 2013;174(17):429-429.

3. Blair J. Bumblefoot: a comparison of clinical presentation and treatment of pododermatitis in rabbits, rodents, and birds. *Vet Clin North Am Exot Anim Pract.* 2013;16(3):715-735.

4. Brown C, Donnelly T. Treatment of pododermatitis in the guinea pig. *Lab Anim.* 2008;37(4):156-157.

5. De Jong IC, Reimert H, Rommers JM. Effect of floor type on footpad injuries in does: a pilot study. *Proc 9th World Rabbit Congr.* 2008.

6. Harcourt-Brown, NH. Foot and leg problems. In: Benyon, PH, Forbes, NA, Harcourt-Brown, NH, eds. *Manual of Raptors, Pigeons and Waterfowl.* Cheltenham, UK: BSAVA; 1996:163-167.

7. Harkness JE, Turner PV, Vande Woude S, Wheler CL. Specific diseases and conditions: pododermatitis. In: Harkness JE, Wagner JE, eds. *Biology and Medicine of Rabbits and Rodents.* 5th Ed. Ames, IA: Wiley-Blackwell; 2010:249-396.

8. Pignon C, Mayer J. Guinea pigs: integumentary disorders. In: Quesenberry KE, Orcutt CJ, Mans C, Carpenter J, eds. *Ferrets, Rabbits, and Rodents: Clinical Medicine and Surgery.* 4th Ed. St Louis, MO: Elsevier; 2020:287-290.

9. Remple JD, Forbes NA. Antibiotic-impregnated polymethyl methacrylate beads in the treatment in the treatment of bumblefoot in raptors. In: Lumeij JT, Remple JD, Redig PT, et al, eds. *Raptor Biomedicine III.* Lake Worth, FL: Zoological Education Network; 2000:255-262.

10. Remple JD. A multifaceted approach to the treatment of bumblefoot in raptors. *J Exot Pet Med.* 2006;15: 49-55.

11. Sanchez-Migallon, D. Birds: Pododermatitis. In: Mayer J, Donnelly TM, eds. *Clinical Veterinary Advisor, Birds and Exotic Pets.* St Louis, MO: Elsevier Health Sciences; 2012:219-221.

12. Varga M. Skin diseases. In: Varga M, ed. *Textbook of Rabbit Medicine.* 2nd Ed. Oxford, UK: Butterworth-Heinemann; 2014:233-240.

13. Varga M, Paterson S. Dermatologic diseases of rabbits. In: Quesenberry KE, Orcutt CJ, Mans C, Carpenter J, eds. *Ferrets, Rabbits, and Rodents: Clinical Medicine and Surgery.* 4th Ed. St Louis, MO: Elsevier; 2020:220-232.

14. Vella D. Rabbits: Pododermatitis. In: Mayer J, Donnelly TM, eds. *Clinical Veterinary Advisor, Birds and Exotic Pets.* St Louis, MO: Elsevier Health Sciences; 2012:407-409.

Echocardiographic Evaluation in Koi Carp (*Cyprinus carpio*) Under Manual Restraint Compared to Anesthesia with Isoeugenol

Megan Partyka, DVM,
Michael Perlini, DVM, Dipl ACVR,
Carl Toborowsky, MS, MA, VMD,
Alvin Camus, DVM, PhD,
Joerg Mayer, DVM, MS, DABVP (Exotic Companion Mammal Practice), DipECZM (Small Mammal), Dipl ACZM

Session #2635

Affiliation: From Department of Small Animal Medicine and Surgery (Partyka, Toborowsky, Mayer), Department of Veterinary Biosciences and Diagnostic Imaging (Perlini), and Department of Pathology (Camus), College of Veterinary Medicine, University of Georgia, Athens, GA 30602, USA.

Note: Abstract format selected by author(s). See Front Matter for description of conference proceedings manuscripts submitted as scientific abstracts.

Abstract: Fish are affected by a wide range of cardiovascular conditions including infectious (e.g., viral, bacterial, fungal, and parasitic) and non-infectious (e.g., congenital, degenerative, environmental, idiopathic) diseases. There is an overall a lack of knowledge and information in the literature regarding both normal cardiac parameters and the cardiovascular effects of anesthetic agents, such as isoeugenol, on fish including koi carp (*Cyprinus carpio*). Koi carp are popular ornamental fish found worldwide as display animals in private collections, zoological institution and aquariums. A cross-over study evaluated echocardiography in 40 adult koi carp and determined ultrasound to be a non-invasive method for cardiac evaluation under both manual restraint and anesthesia with isoeugenol (AQUI-S) at 50 ppm. Mean heart rate was significantly higher in anesthetized fish compared to manually restrained fish. Respiratory rate decreased significantly in anesthetized fish compared to manually restrained fish. Mean ejection fraction, stroke volume, and cardiac output were significantly lower in anesthetized fish than in manually restrained fish. Overall poor measurement agreement and correlation were found between veterinarians measuring the same echocardiography study for each fish. This study provides the first description of echocardiography in koi carp and demonstrates significant differences in cardiopulmonary parameters under manual restraint compared to anesthesia with isoeugenol.

Clinical Findings and Surgical Repair of Acquired Coelomic Hernias with Polypropylene and Porcine Mucosa Meshes in a White Spotted Bamboo Shark (*Chiloscyllium plagiosum*)

Julie Pujol, DMV,
Claire Vergneau-Grosset, DMV, IPSAV, CES, Dipl ACZM,
Marion Jalenques, DMV,
Juliette Raulic, DMV, IPSAV,
Marie-Pierre Lessard,
Wahiba Ait Youcef,
Noémie Summa, DMV, IPSAV, Dipl ACZM

Session #2711

Affiliation: From the Faculté de médecine vétérinaire, Université de Montréal, 3200 rue Sicotte, Saint-Hyacinthe, QC J2S 2M2, Canada (Pujol, Vergneau-Grosset, Jalenques, Raulic, Summa), The Aquarium du Québec, Sépaq, 1675 avenue des hôtels, QC G1W 4S3, Canada (Lessard, Ait Youcef).

Note: Abstract format selected by author(s). See Front Matter for description of conference proceedings manuscripts submitted as scientific abstracts.

Abstract: A captive adult male white spotted bamboo shark exhibited a sudden onset of depressible distension of the cranial third of the coelom. A discontinuity of the coelomic muscle wall associated with moderate coelomic effusion was noted on coelomic ultrasound. Coelomic hernia and focal coelomitis were suspected and treatment was initiated (ceftiofur 6.6 mg/kg and florfenicol 40 mg/kg, IM once). Hematology and biochemistry revealed an increase in urea and total proteins.[1,2] Upon exploratory coeliotomy under anesthesia (MS-222 50 mg/L and propofol 2.5 mg/kg IV[3]), a double coelomic hernia of the ventral coelomic wall without entrapment of vital structures was confirmed. The larger hernia of the transversus abdominis[4] muscle was reduced with a polypropylene mesh (Prolene mesh, Ethicon, Inc.) and a porcine submucosa mesh implant (Surgisis, Cook Medical, Ind.). The second hernia was closed with an interrupted cruciate pattern. Bacterial culture of the muscular wall revealed a multidrug resistant *Pseudomonas* sp. Postoperative treatment included gentamycin 6 mg/kg IM q24h × 7 days based on culture sensitivity, meloxicam 0.5 mg/kg once and nutritional support. Postoperative complications included skin erythema and subcutaneous edema adjacent to the wound for up to 1 month post-surgery. Inflammatory nodules developed on the scar four months after the herniorrhaphy. No wound dehiscence or hernia recurrence were noted on follow-up 9 months post surgery. The definitive etiology of the hernia remains unknown but it is suspected to be traumatic. To our knowledge, this is the first report of a successful treatment of coelomic hernia with polypropylene and porcine mucosa meshes in elasmobranchs.

References

1. Alexander AB, Parkinson LA, Grant KR, et al. The hemic response of white-spotted bamboo sharks (*Chiloscyllium plagiosum*) with inflammatory disease. *Zoo Biol.* 2016;35(3):251-259.

2. Parkinson LA, Alexander AB, Campbell TW. Variability in hematology of white-spotted bamboo sharks (*Chiloscyllium plagiosum*) in different living environments. *Zoo Biol.* 2017;36(4):284-288.

3. Miller SM, Mitchell MA, Heatley JJ, et al. Clinical and cardiorespiratory effects of propofol in the spotted bamboo shark (*Chylloscyllium plagiosum*). *J Zoo Wildl Med.* 2005;36(4):673-676.

4. Ashley LM, Chiasson RB. *Laboratory Anatomy of the Shark.* 5th Ed. Ontario, Canada: McGraw-Hill College; 1988.

Point-of-Care Ultrasonography in Exotic Animals

Cathy A. Johnson-Delaney, DVM

Session #2733

Affiliation: From Washington Ferret Rescue & Shelter, 12522 128th Lane NE, Kirkland, WA 98034, USA, NW Zoological Supply, 3614 Smith Ave, Everett, WA 98201, USA.

Abstract: Point-of-care ultrasound (POCUS) and AFAST (abdominal focused assessment with sonography for trauma), and TFAST (thoracic focused assessment with sonography for trauma) techniques can be done in exotic patients such the ferret, rabbit, guinea pig, parrot, chicken, bearded dragon, chameleon, chelonian, and snakes. This POCUS can quickly look for fluids in the body cavities and be used as screening for other diseases such as masses, organomegaly, and uroliths. Sonographic uses for aspirations, and needle biopsies will be discussed. POCUS to assess the heart can facilitate initiation of treatment, although it may not take the place of a detailed echocardiograph. Ultrasound is also useful to visualize gut motility. Because of differences in anatomy, scanning positions for different organs are used. In rabbits and guinea pigs, paralumbar scanning is useful for kidneys and the reproductive tract. Ferret positioning differs from those used in dogs and cats. Assessment in reptiles requires anatomic knowledge but is useful for coelomic enlargements. A standoff may be necessary for scanning reptiles and birds. The handheld scanner that projects onto a cellphone or tablet computer makes point-of-care scanning practical to add to any examination.

Introduction

Wouldn't we all love to be able to use a Star Trek medical tricorder and scanner right now? In some ways, use of a point-of-care ultrasound (POCUS) device, particularly one that is a handheld unit going wirelessly to a cell phone, tablet or computer gives us some of that ability right now! While doing a physical examination is still a starting point for the hands-on exam, results can be inconsistent if you encounter a tense body, even after administration of a sedative. In many, you cannot fully palpate all the organs because other organs are in the way, such as a cecum in a rabbit or guinea pig making palpation of the rest of the abdomen difficult. Being able to observe organs, fluids, and even function (think heart) while having the animal on your exam table or cageside, can expedite care timelines, improve outcomes, and even give clients peace of mind that you are really looking thoroughly at their pet. Ultrasound units are now in a price range and of improved technology that every practice can afford to have it available.

An animal needs to be in an appropriate position to be comfortable, while allowing you to scan through the abdomen's four quadrants. Consider the cranial quadrant containing the liver, spleen, stomach and associated lymph nodes. The left quadrant contains the left kidney, adrenal, retroperitoneal space. From the same probe position, you can scan the caudal abdomen including the urinary bladder and reproductive tract. The right quadrant contains the right kidney adrenal, pancreas, and associated lymph nodes of the mesentery. In rabbits and rodents with large cecums, you may need to move the probe ventrally to look at the bladder and reproductive tracts. But even so, within 3-5 minutes you can have some answers – this essentially is the abdominal focused assessment with sonography for trauma (AFAST) technique so you also will be able to tell if there is free fluid in the abdomen. You can apply the thoracic focused assessment with sonography for trauma (TFAST) technique to the thoracic area, and quickly look for free fluid, major cardiac dysfunction, even air or fluid within the lungs. Importantly for many species, the ability to scan the cranial thoracic area can rule out masses versus enlarged lymph nodes.[1-8]

Application for Use in Exotic Pets

POCUS techniques have been described in detail for dogs and cats, but unfortunately applications for use in exotic pets have not been published. It is possible to apply these techniques, but given the varied anatomy of the different mammals, birds and herpetologic species, the practitioner requires guidelines for the best positioning of the probe. Practitioners need to: (1) have knowledge of the normal anatomy of the different species and (2) practice! Do a lot of scans on normal, healthy individuals of the different species. In small animals, and particularly in those with little cutaneous fat, a stand-off may be needed. Herpetologic species usually need adjustment to account for the presence of scales, scanning from the ventrum where the scales may be thinner can help. In chelonians, you image through leg fossa. The size of the probe versus the size of the chelonian may limit what you can do, although using a standoff in the small leg fossae will allow you to image a fair amount of the internal body. These can be made using distilled water or ultrasound gel filled procedure glove. There are commercial standoffs available, but most are too large for the animals you need to scan. (3) Practice! Do scans on ill animals. Many of these should be following administration of an analgesic and/or a mild sedative. If an animal is being anesthetized for radiographs or other diagnostics, while the animal is being induced or recovered, you can quickly check some organs or body cavities. For example, every male ferret over the age of 3 years, quickly scan the bladder and prostate during the physical examination. POCUS can be incorporated into every geriatric animal examination. The examination fee can be adjusted to just include this 3- to 5-minute scan.

Remember that gas/air, bone, and hair can introduce artifacts. Uroliths or cecoliths or even calcified eggs, tumors or other structures will cause artifacts and difficulties imaging organs that are masked by these. This is why you can tailor POCUS to any animal and look at almost any organ. Birds, however, are still difficult due to air sacs, but by using an organ such as the liver as a window you can image the heart. The avian GI and reproductive system can be challenging, but if you locate the ventriculus, you can actually see the contractions, and through it you may be able to see gonads, the spleen, and even assess the proventriculus. You just need to picture the anatomy and move the probe where it can penetrate to the organ or body cavity to be viewed. Again, many of us learned ultrasound as a rather static imaging system with specified probe locations and angles, but this was in dogs and cats. Be creative with how you do imaging. It is also helpful to image an animal before going to surgery or necropsy, and then compare the ultrasonographic image to what is actually present. Again, practice!

It is advisable to have ready resources showing the normal anatomy of various species. There are some references that have enough ultrasound pictures of normal anatomy to aid in viewing any patient.[9,10] Examples of scans that can be of immediate help on the examination table or cageside include the following.

Birds

Abdominal area, midline: presence of egg, fluid, stasis of ventriculus, masses, liver disease. Angle the probe through the liver cranially to view the heart. Fluid in the pericardium and general status of the heart.

Lizards

Paralumbar, lumbar fossa area to view kidneys, from ventrum: bladder, liver, reproductive tract. Probe placement axillary area aiming towards shoulder girdle, imaging of the heart.

Snakes

Ventrum, using a standoff: view entire snake. Especially helpful to look for obstructions, masses, retained eggs, cecoliths.

Chelonians

All leg fossas: internal anatomy.

Ferrets

Usually scan with the ferret either in dorsal recumbency or scruffed and held up. Start at bladder, prostate. Doing palpation along with the ultrasound can visualize quickly almost all organs. Angling up cranially through the liver may show fluid in the thoracic cavity. The heart and cranial mediastinum are easily viewed.

Rabbits, guinea pigs, and others with large cecums

From para/lumbar fossa area view kidneys, and possibly some of the reproductive tract. From the caudal ventrum view the bladder directly and use the bladder as a standoff to view the reproductive tract. The liver is viewed moving the probe ventrally and tracing the liver silhouette. The heart is imaged from the lateral thoracic areas, realizing rib artifact. The cranial mediastinum can be imaged as well from the lateral body wall.

Other small animals

Hedgehogs: sedation is almost always needed to be able to image. All images come from under the mantle. In lateral recumbency, image positions are close to those described for a dog.

Rodents

The probe size may image almost all internal structures from the ventrum of the abdomen or the lumbar fossa areas. Hamster cheek pouches can be checked by imaging over the pouch.

Conclusions

Principles of POCUS can be used in exotic pet practice, although positioning of the probe may be different depending on species. The practitioner needs to be familiar with normal anatomy, as well as be able to recognize abnormalities. The development of handheld or portable ultrasound probes makes it easy for all exotic practitioners to do ultrasonography. It also has become economically feasible to own the equipment.

References

1. Boysen SR, Rozanski EA, Tidwell AS, et al. Evaluation of a focused assessment with sonography for trauma protocol to detect free abdominal fluid in dogs involved in motor vehicle accidents. *J Am Vet Med Assoc.* 2004;225(8):1198-1204.

2. FASTVET. Available at: http://www.FASTvet.com. Accessed April 2, 2022.

3. Lisciandro GR, Lisciandro SC. Global FAST for patient monitoring and staging in dogs and cats. *Vet Clin North Am Small Anim Pract.* 2021;51:1315-1333.

4. Lisciandro GR. AFAST target-organ approach and fluid scoring system in dogs and cats. *Vet Clin North Am Small Anim Pract.* 2021;51:1217-1231.

5. McMurray J, Boysen S, Chalhoub S. Focused assessment with sonography in nontraumatized dogs and cats in the emergency and critical care setting. *J Vet Emerg Crit Care.* 2016;26(1):64-73.

6. Lisciandro GR. Abdominal and thoracic focused assessment with sonography for trauma, triage and monitoring in small animals. *J Vet Emerg Crit Care.* 2011;21(2):104-22.

7. Boysen SR, Lisciandro GR. The use of ultrasound in the emergency room (AFAST and TFAST). *Vet Clin North Am Small Anim Pract.* 2013;43(4):773-797.

8. Lisciandro GR. *Point-of-Care Ultrasound Techniques for the Small Animal Practitioner.* 2nd Ed. Ames, IA: Wiley-Blackwell; 2021.

9. Krautwald-Junghanns ME, Pees M, Reese S, Tully R. *Diagnostic Imaging of Exotic Pets.* Hannover, Germany: Schlutersche Verlagsgesellschaft mbH & Co; 2011.

10. Johnson-Delaney CA. Ultrasonography in ferret practice. In: Lewington JH, ed. *Ferret Husbandry, Medicine and Surgery.* 2nd Ed. Sydney, NSW, Australia: Elsevier Saunders; 2007:417-429.

Exotic Emergency Triage, Medicine, and Surgery

Meera Music, BVM&S, MRCVS,
Alicia McLaughlin, DVM, CertAqVet,
Anneliese Strunk, DVM, DABVP (Avian Practice)

Session #2569

Affiliation: From The Center for Bird and Exotic Animal Medicine, 11401 NE 195th St. Bothell, WA 98011, USA.

Abstract: Exotic animals commonly present to veterinary clinics on an emergency basis. Both acute and chronic diseases may manifest with a sudden onset of symptoms due to the tendency of exotic species to hide signs of illness until it is advanced. Accurate diagnosis of the underlying cause for illness is essential to provide appropriate treatment, but in many cases, diagnostic testing must be staged to avoid inadvertent mortality of these fragile patients. In this presentation, we discuss how to optimize the clinical approach to the critically ill exotic animal, and review emergency triage, medical, and surgical management of exotic mammals, reptiles, and birds.

Introduction to Triage

The purpose of this presentation is to aid the emergency (ER) clinician in their ability to understand and utilize accurate triage to manage a hectic (typical) ER shift. Appropriate triage is one of the most valuable tools in any ER clinician's repertoire, and can improve efficiency, optimize staff and resources, and advance patient and client care.[1] Speed and accuracy during triage are essential on an ER shift, as the condition of patients can rapidly change.

Exotic pets frequently arrive at emergency veterinary clinics in critical condition and may be suffering from one or more concurrent chronic diseases resulting in acute presentation.[2] There is a large body of literature reviewing emergency medical techniques in exotic animals that clinicians should familiarize themselves with prior to treating these patients. Emergency techniques from dog and cat medicine can often be extrapolated to exotic patients, although anatomy and physiology will vary widely depending on species.

Triage is an essential skill for both the ER clinician and other veterinary support staff. Triage must occur numerous times throughout the intake process including the following:

1. Scheduling: reception staff communicates with the client and establishes whether the reported symptoms warrant an emergency visit or can be scheduled as an urgent care appointment or routine exam

2. Entrance of the animal to the emergency clinic: veterinary technicians or assistants acquire a brief history and perform an initial visual assessment

3. Physical exam: the veterinarian examines the patient and orders stabilization care (when/if appropriate)

4. Diagnostic sample collection/procedures: staff and resources must be organized to determine order of diagnostic sample collection, placement of intravenous (IV) or intraosseous (IO) catheters, setting up continuous rate infusions (CRIs), and prioritization/timing of indicated procedures

5. Diagnosis: interpretation of the onsite diagnostic results and communication of the findings with the client

6. Treatment plan: prioritizing order of additional treatments/procedures based on test results

Standardization of triage is important to consistently and accurately communicate the condition of the animal, both to the client and to fellow staff members. This also aids in scheduling and sign-posting an accurate timeline to clients. Managing client expectations is an important part of an ER day, and keeping communication open is key.

Triage lists can be an excellent way to standardize patient care (Table 1). They can assist receptionists in making scheduling recommendations and help medical team members accurately prioritize critical patients. Any time phone triage occurs, it is important to warn clients of likely wait times based on the current status of the ER caseload. Clients should also be cautioned that ultimately, an in-person exam combined with diagnostic evaluation is the only way to determine accurate triage status.

Clinical symptoms of exotic pets are often masked at home as part of their drive to hide illness.[3] Symptoms are often hidden even more effectively in clinical settings secondary to stress, especially on initial presentation. Allowing a patient time to adjust to a new environment may aid in clinical assessment; however, if an animal is in critical condition, rapid decompensation is possible anytime from presentation to during or after assessment, as the ultimate presentation of acute-on-chronic disease.[4]

On presentation at the clinic, the technician or assistant should gather a brief history at the door to help prioritize patient care and identify the particular client's concerns. Important information that must be covered includes the following:

1. Whether or not the owner wishes for cardiopulmonary resuscitation (CPR) efforts to be pursued in the event of cardiopulmonary arrest.

   This should be noted clearly in the medical chart and on the patient's enclosure with either CPR or DNR (do not resuscitate) to improve clinic response in the event of an emergency. All CPR patients should have dosages of emergency drugs calculated at the time of presentation and kept with the patient at all times to maximize the chances of a successful CPR. There are automatic drug calculators that can be used to easily and quickly make these calculations.[5,6]

2. Client permission to provide supportive care if the animal appears to need pain medication or fluid support.

   If the animal is actively bleeding or struggling to breathe, then the only conversation that should be had prior to bringing the animal to the treatment area should be CPR status to prevent delays in providing appropriate patient stabilization.

A more detailed medical history, including husbandry conditions, can be triaged to a later time. Alternatively, stock electronic questionnaires (such as JotForms.com) can be utilized so the client can fill this out while their animal is being initially assessed.

Table 1. Triage system.

Color: Urgency (time frame)	Description of animal condition
Red: Now	Severe lethargy, birds sitting on the bottom of enclosure and fluffed Any animals laying on side/unable to stand Seizure Severe trauma, even if the animal appears to be ok Uncontrolled bleeding Fractures with open wounds or protruding bone Difficult, rapid, or labored breathing Not eating for 24+ hours (not reptiles) Animals that have been scratched/bitten by a cat, mammal, or bird Painful postures, hunched over, possibly rigid, possibly grinding teeth Genital bleeding, exhibiting signs of a prolapse Very small or young bird or mammal not eating for 12+ hours.
Orange: Today	Diarrhea Vomiting Poor response to stimulation Mild lethargy Birds, not acting quite right Diminished stool production over last day or days Concerns about recent surgery Any open wounds Eye injuries
Yellow: Tomorrow	Sneezing Eye discharge Masses Ear infections
Green: Can be seen in the next few days	Reptiles that are inappetent or constipated Reptiles with bad sheds Mild skin problems with no other concerns Annual or boarding exams with no other concerns

Emergency patients will be seen and given immediate stabilization treatment in order of triage severity. Unfortunately, exotic pets tend to hide symptoms of illness until they are so sick they cannot do so any longer, and they can suddenly decompensation without warning. Triage status may change as more information is obtained and as exam findings change. Wait times vary depending on caseload and may exceed 6 hours.

After the patient arrives at the clinic, the doctor should perform an immediate visual examination of the animal. This can occur even if the vet is on the phone with another client. If the animal is in critical condition or in need of more direct support, it is appropriate for the veterinarian to tell the client on the phone that they must end the conversation to tend to an emergency that has just come through the door.

Visual examination should include evaluating mentation, posture, and respiratory rate/effort. If the animal is struggling to breathe on presentation, it should be immediately placed in an oxygen chamber prior to a hands-on physical exam to permit stabilization. Ideally, the reception team should inform the treatment team that a dyspneic animal emergency was scheduled, so that an O_2 chamber can be prepared ahead of presentation.[3,7]

A temperature should be taken from mammals on intake (when feasible - this may not be possible for species such as hedgehogs). Rectal temperatures are an objective measurement of stability of the patient. Prognosis for rabbits[8] and guinea pigs[9] is significantly reduced when these animals are hypothermic on presentation. As temperatures continue to decrease below intake rectal temperature, the odds of death significantly increase. Another immediate point-of-care analysis that should be available is blood glucose. This is particularly relevant in the classic case of a seizuring or apparently comatose ferret secondary to an insulinoma, but it is also helpful in neonatal or very small animals. These animals are often hypothermic on presentation, but providing heat alone without glucose supplementation and complete, balanced nutritional support could result in a secondary crash after the patient is normothermic. Be aware that most point-of-care glucometers are less accurate compared to i-stat or VetScan analyzers for exotic species.[10]

Exams on critical patients should be performed in a staged fashion to minimize the length of time the patient is experiencing stress.[2] For example, the most prudent approach to an exam of a dyspneic cockatiel would be to quickly assess the following parameters while transferring the animal to oxygen: a brief palpation of the keel to determine body condition, which can indicate chronicity of the issue; brief coelomic palpation, and brief auscultation, which can alert the clinician to presence of coelomic or other disease.[3] The patient should then be placed in an oxygen chamber while a conversation with the client occurs that outlines realistic possible outcomes at this stage, including death. A delicate patient may pass away even with a cursory exam, let alone one that requires further handling, and so warning the client of possible outcomes prior to embarking on a more thorough examination is the kindest, and most appropriate option.

Once a complete examination has been performed, a follow up conversation with the client must occur, outlining the physical exam findings in more detail, the differentials that could be involved in the possible disease processes, possible prognosis, and diagnostic testing recommendations. Discussion of euthanasia in a case where the animal is in critical condition (especially when the client has expressed financial constraints) is important. Though it can be difficult to discuss ending the life of a family member, it is also important not to be vague or ambiguous about costs of long-term care (financial and time/effort) so that the client is able to make a fully informed decision. Even if finances do not play a significant role in medical decisions for the client, trying to be realistic about prognosis is very important so clients are fully aware of likely outcomes is very important.

Once the decision to move forward with hospitalization is made between the client and veterinarian, and the risks, prognosis, and finances have been discussed, then it is time to decide how to stage the diagnostics and treatments. The client should understand that the treatment plan is based on the patient's response to therapy, and/or changes to repeat diagnostic testing, as this will directly impact the decision to further hospitalize, move to surgery, or consider euthanasia.

It is important to prioritize diagnostics in order of those that will change the treatment approach. For example, in the case of the dyspneic patient, both radiographs and a complete blood count (CBC) would be important diagnostics to consider, but if finances were a concern, or staging the diagnostic tests needed to occur due to patient instability, radiographs would give more immediately actionable clinical information if the animal has evidence of pneumonia versus evidence of perihilar edema and an enlarged heart which is indicative of heart disease. [7] The decision to place a dyspneic animal on antibiotics with evidence of pneumonia on radiographs does not require an elevated white blood cell count or relative heterophilia on a CBC to make that clinical decision; however, these additional diagnostic tests still have value, as they may reveal unexpected pathologic changes, provide additional prognostic information, and provide the clinician with the ability to monitor trends and response to therapy.[3]

One useful tool when embarking on an exam of a dyspneic, stressed, or painful patient that is at risk for decompensation is sedation. Many sedation protocols exist, though at the authors' hospital low doses of midazolam ± butorphanol or buprenorphine are commonly utilized to lower stress levels and relax the patient. This can often improve outcomes when needing to handle the patient for any period of time out of the comfort of an oxygen chamber.[4]

Keep in mind with dyspneic avian and reptile patients, positioning can affect the stress level of the patient. In birds, an upright posture is ideal to reduce pressure on the respiratory tract. In the event that radiographs, bloodwork, and ultrasound are all approved for an avian patient with coelomic distention and/or respiratory distress, the first diagnostic to perform should be the ultrasound. Coelomic ultrasound can provide important information about the level of coelomic disease present.[3] If fluid is identified on ultrasound, it should be aspirated prior to taking radiographs, as reduction of the fluid will help to improve radiographic coelomic detail and will also drastically improve the animal's comfort, ability to breathe, and stability.[11] Remember that unpositioned dorsoventral and horizontal beam lateral radiographs can be used so that the patient can be kept in a normal, anatomic, upright posture, rather than restraint that may increase dyspnea. The final step in the diagnostic workup of a dyspneic animal is often phlebotomy, and this may be done immediately after radiographs. If the patient appears stressed or declining due to the positioning needed to obtain the sample, then the animal should be placed back into oxygen for a break prior to phlebotomy.

In the event ultrasound had revealed a significant problem (large mass present, large volume of fluid), then prior to performing the rest of the approved diagnostics, it would be best to discuss this current finding with the client. It is important for the client to be included in the decisions of the animal's care – even if the mass type or source of fluid is unknown, what is known is that it is significant enough to affect breathing to the point that the animal cannot breathe comfortably out of oxygen. It would be important to have another conversation regarding prognosis, possible medical therapeutic options (for example, in the case of reproductive disease leading to fluid accumulation or tumors, GNRH agonist therapy may be an excellent therapeutic option), surgical options, ongoing oncology referral, or euthanasia.[12] In the event that surgery is desired, then determining what is necessary for the animal to stabilize to the point the surgery is feasible is the next consideration.

If the client is interested in continuing treatment, then the rest of the diagnostics should be gathered to obtain information regarding the animal's stability, organ function, among other things, because if there are concurrent pathologic changes, they need to be addressed prior to surgery. It can be tempting when an animal presents in an emergency to have a more fundamental approach to a mass: cut it out! But it is important to go in a stepwise manner through the possible outcomes, and be realistic and patient when needed. Although surgery is a vitally important tool in any ER vet's repertoire, time needs to be taken to determine if surgery is truly in the best interest of the patient and if it is, then the stage should be set to give the patient the best possible outcome during and after surgery.

Surgery

Surgical intervention is required in a number of exotic emergency cases;[13] however, stabilization and careful preparation is required to maximize the chance of successful outcomes. In many cases, administration of fluid therapy, analgesia (including constant rate infusions), oxygen support (if indicated), control of hemorrhage via short-term application of pressure wraps, and time will improve an emergency patient's chances of surviving an anesthetic procedure.

Select cases that require immediate surgical intervention

The following are examples of cases that require immediate surgical intervention:

- Uncontrollable hemorrhage (no response to direct pressure, inability to place a pressure wrap)
- High likelihood of internal hemorrhage or damage
  - Liver lobe torsions with free fluid, decreasing packed cell volume (PCV), or uncontrollable pain
  - Extensive prolapse of organs from the cloaca (birds or reptiles) or rectal prolapse in mammals
  - Penetrating injuries to the thorax, abdomen, or coelom (predator bite wounds, projectile injuries)
- Tracheal obstruction (requires tracheostomy in mammals or airsac cannulation in birds)[14]
- Egg retention with respiratory distress that does not respond to oxygen therapy
- Severe abdominal pain that does not respond to analgesic therapy (including CRI administration)
- Sudden temperature drop and/or change in attitude in mammals with GI signs
- Herniation

In an effort to improve successful surgical outcomes, preparation is critical, even in emergency situations. Staff should be trained to place IV catheters, initiate CRI administration, and perform blood transfusions (especially autotransfusions) quickly. Ensure that equipment and tools are labeled and available for easy access. Utilize procedure equipment checklists, drug calculators, and walkthroughs with staff to prevent key supplies from being forgotten. There are resources to help with setting up protocols for constant rate infusions.[15,16]

Common emergency procedures performed at CBEAM

The following are emergency procedures that are commonly performed at CBEAM:

- Trauma surgery-laceration repairs, wound exploration, enucleations, amputations, fracture repairs, etc.
- Occlusal adjustments when intraoral severe soft tissue trauma is present (exotic mammals)
- Avian tracheoscopy and airsac cannulation

- Ferret urethral catheterization (with possible stone retropulsion)

- Ferret GI obstruction
  - Gastrotomy
  - Enterotomy
  - Intestinal resection and anastomosis

- Rabbit liver lobectomy (with concurrent autotransfusion if hemorrhage present)

- Gastric decompression

- Rabbit abdominal exploration
  - Gastrotomy
  - Enterotomy

- Reptile coelomic exploratory surgery/foreign body removal

Surgical intervention is not without risk, and post-operative recovery can be complicated and expensive. For these reasons, detailed discussions with owners are required even in emergency cases, so they are prepared for possible negative outcomes, and for the long-term expectations for case management and cost.

For example, a bird in respiratory distress with a tracheal obstruction would improve with placement of an airsac cannula, but if an owner is unable or unwilling to proceed with the diagnostics to identify the problem or treatments to correct the problem (that might include surgical intervention or months of oral ± nebulized medication), euthanasia should be recommended as an option from the start. If an owner is not sure about long term commitment, it still may be indicated to place an airsac cannula to allow the owner time to process the situation. It should be made clear that placement of the tube is not an insignificant procedure and requires hospitalization while the tube is in place.

Some procedures have historically been considered too dangerous to perform in exotic species, either related to complications from the procedure or anesthesia. With appropriate anesthetic monitoring (by both equipment and staff), maintenance of an airway/ventilatory support, and fluid support with IV or IO catheterization, anesthetic safety has dramatically improved in exotic species. Rabbit/rodent abdominal exploration and GI surgery has been considered too likely to result in post-operative complications in the past; however, if an animal is showing clear indications for surgical intervention, especially related to an obstructive process in the GI tract, the best outcomes are achieved from addressing the problem sooner than later to prevent issues from necrosis the longer an obstruction persists. In addition, this will minimize the amount of suffering the patient experiences if it turns out that the problem is beyond surgical correction, and euthanasia on the table may be the most compassionate decision. Please see below for further discussion of rabbit/small mammal GI disease surgical management.

Specific case examples

Severe avian trauma

In the author's (AS) experience, birds that sustain a traumatic partial amputation from a predator attack tend to have a better outcome when stabilized for 24 to 48 hours before surgery. As long as there is no active bleeding and no coelomic penetration has occurred, birds should be bandaged and given time to recover from the shock of the traumatic event before surgery.

Small mammal gastric dilation

Initial stabilization of cases with moderate to severe gastric dilation should include intensive management with IV fluids, CRIs, and anxiolytics to improve patient comfort. This may help symptoms resolve without the need for further intervention (such as gastric decompression or exploratory surgery). Gastric decompression is a technique that should be performed promptly if there is no improvement with initiation of the intensive care, and should be performed before surgical intervention is recommended.[17] Decompression can rapidly decrease the amount of tension within the gastric wall (which can lead to necrosis), dramatically increase patient comfort, and alleviate pressure on sections of the intestinal tract that might be compressed by the dilated stomach.

Decompression should be performed under general anesthesia using a large red rubber catheter that has been fenestrated and lubricated prior to passing into the stomach.[18] Severely depressed patients may only need sedation for decompression. Risks of the procedure may include rupture of the cranial GI tract, aspiration, and destabilization secondary to anesthesia. If blood is seen on the red rubber catheter, then the procedure should not be continued or repeated. Blood in this instance does not necessarily indicate perforation of the GI tract but does suggest severe mucosal irritation or inflammation.

If the gastric dilation is easily alleviated with decompression, but recurs afterwards, more than one decompression may be indicated; however, this makes it more likely there is another problem in the distal GI tract. If the gastric dilation continues to recur rapidly after decompression, or if the stomach cannot be easily/effectively decompressed, an exploratory surgery/gastrotomy should be performed.

Liver lobe torsion

Liver lobe torsions (LLT) have been identified in rabbits presenting with a wide variety of clinical signs. This includes rabbit GI stasis syndrome (RGISS) or any other nonspecific clinical signs of illness.

Physical exam findings of LLT may include cranial abdominal pain, palpation of a firm liver lobe, decreased body temperature, and pale mucous membranes; however, many rabbits present with sudden onset of inappetence as the only symptom. The authors have not noted a clear breed or sex predisposition, although lops have been previously reported to be overrepresented.[19] Classic laboratory findings supportive of LLT include anemia (usually moderate, sometimes severe) and mild to severe elevations in alanine transaminase (ALT). The range of ALT values documented in confirmed LLT is 89 mg/dl to >1000 mg/dl. Radiographic findings are inconsistent.

Although LLTs can be diagnosed on computed tomography (CT),[20] ultrasound is the most useful and rapid diagnostic modality for identification of the affected lobe(s). It is vital to visualize all lobes and compare blood flow between each lobe to accurately identify an absence or relative decrease in blood flow. The caudate lobe is the most frequently affected, but other liver lobes can also torse, and more than one lobe may be affected in the same animal. To identify the caudate lobe most easily, the authors utilize the following approach. Place the ultrasound probe caudal to the right side of the ribcage aiming dorsally. Visualize the right kidney and sweep cranially to identify the caudate lobe, which is usually in contact with the kidney. Evaluate the echotexture in comparison to the other lobes and use Doppler to assess blood flow. If free fluid is present, be prepared for performing an autotransfusion during surgery.

Although some rabbits will survive with a torsed liver lobe, surgical excision of the affected lobe(s) results in the fastest resolution of clinical signs and ongoing risk. With practice and an assistant surgeon to help with retraction and visualization, this procedure can be performed in less than 20 minutes. If the PCV is less than 20%, blood transfusion should be considered regardless of whether surgery is elected. The authors have seen two cases of concurrent gastric torsion in rabbits with LLT.

Advanced Imaging in Emergency Settings

CT is an advanced imaging modality that can be used in an emergency setting to rapidly identify problems, better determine prognosis, and guide treatment recommendations. In many cases, quick scans can be performed with sedated patients securely wrapped in a towel and/or placed in a plastic container. More intensive scans with contrast can be considered depending on patient stability and vascular access.

Case presentations that would benefit from CT evaluation

Cases that benefit from CT evaluation include the following:

- Abdominal wall hernias
- Suspected LLT
- Penetrating injury/traumatic injury
- Dyspnea/lower airway disease
- Nasal hemorrhage/obstruction
- Coelomic/abdominal disease
- Appendicitis in rabbits[21]

Summary and Conclusions

Addressing and standardizing triage across every aspect of the intake procedure of a patient is an important part of effective ER management. By appropriate triage at the outset of patient presentation and reevaluating case status numerous times throughout the hospitalization process, an ER clinician can make the best medical, diagnostic, and surgical recommendations, significantly improving patient outcomes.

Examples of emergency cases seen by an exotic emergency veterinarian are provided for discussion in Appendix 1.

References

1. Mathews K. *Veterinary Emergency and Critical Care Manual.* 3rd Ed. Guelph, ON: LifeLearn; 2017.
2. Graham J, Doss G, Beaufrere H. *Exotic Animal Emergency and Critical Care Medicine.* Hoboken, NJ: Wiley-Blackwell; 2021.
3. Music M, Strunk A. Reptile critical care and common emergencies. *Vet Clin North Am Exot Anim Pract.* 2016;19(2):591-612.
4. McLaughlin A, Music M, Strunk A. Stayin' alive: current topics in exotic emergency medicine. *Proc ExoticsCon.* 2019;20-33.

5. Costello M. Principles of cardiopulmonary cerebral resuscitation in special species. *Sem Avian Exot Pet Med.* 2004;13(2):132-141.

6. Lichtenberger M. Shock and cardiopulmonary-cerebral resuscitation in small mammals and birds. *Vet Clin North Am Exot Anim Pract.* 2007;10:275-291.

7. McLaughlin A, Strunk A. Common emergencies in small rodents, hedgehogs, and sugar gliders. *Vet Clin North Am Exot Anim Pract.* 2016;19(2):465-499.

8. Di Girolamo N, et al. Rectal temperature is a prognostic indicator in client-owned guinea pigs. *J Small Anim Pract.* 2021;62(10):861-865.

9. Di Girolamo N, Toth G, Selleri P. Prognostic value of rectal temperature at hospital admission in rabbits. *J Am Vet Med Assoc.* 2016;248(3):288-297.

10. Petritz O, Antinoff N, Chen S, et al. Evaluation of portable blood glucose meters for measurement of blood glucose concentration in ferrets (*Mustela putorius furo*). *J Am Vet Med Assoc.* 2013;242(3):350-354.

11. LafeberVet. Abdominocentesis in birds. Available at: https://lafeber.com/vet/abdominocentesis-in-birds/. Accessed Feb 1, 2022.

12. LafeberVet. Reproductive emergencies in birds. Available at: https://lafeber.com/vet/reproductive-emergencies/. Accessed Feb 1, 2022.

13. Bennett RA, Pye GW. *Surgery of Exotic Animals.* Hoboken, NJ: Wiley-Blackwell; 2022.

14. LafeberVet. Air sac cannula placement. Available at: https://lafeber.com/vet/air-sac-cannula/. Accessed May 27, 2022.

15. Schnellbacher R, Comolli J. Constant rate infusions in exotic animals. *J Exot Pet Med.* 2020;35:50-57.

16. Veterinary Information Network. Constant rate infusion calculator. Available at: www.vin.com. Accessed Feb 1, 2022.

17. Lennox A. Updates in medical versus surgical management of rabbit GI stasis syndrome. *Proc ExoticsCon.* 2021.

18. Guerrera B. Successful decompression technique in rabbits presenting with gastric outflow obstructions. *Proc ExoticsCon.* 2019;423-426.

19. Graham JE, Orcutt CJ, Casale SA, et al. Liver lobe torsion in rabbits: 16 Cases (2007 to 2012). *J Exot Pet Med.* 2014;23(3):258-265.

20. Dagget A, Loeber S, Le Roux A, et al. Computed tomography with Hounsfield unit assessment is useful in the diagnosis of liver lobe torsion in pet rabbits (*Oryctolagus cuniculi*). *Vet Radiol Ultrasound.* 2021;62(2):210-217.

21. Di Girolamo N, Petrini D, Szabo Z, et al. Clinical, surgical, and pathological findings in client-owned rabbits with histologically confirmed appendicitis: 19 cases (2015-2019). *J Am Vet Med Assoc.* 2021;260(1):82-93.

Appendix 1. A day in the life of an exotic emergency veterinarian. Cases for discussion.

1. The case of the fluidy fowl
   - Owner brought chicken in for suspected egg binding and lethargic behavior
   - Physical exam shows severe ascites with no palpable egg(s)

2. The case of the steamed Senegal
   - Parrot landed in a pan of boiling water
   - Mild inflammation at burn site

3. The case of the flying ferret
   - Ferret jumped off a balcony
   - Femoral fracture palpable, very painful, hematoma at site

4. The case of the blah bunny
   - Normal temperature, mildly dehydrated, quiet, tense abdomen

5. The case of the lethargic lagomorph
   - Low body temperature (98 °F), mildly dehydrated, quiet, slightly pale mucous membranes

6. The case of the wryneck rabbit
   - Rabbit with sudden onset of head tilt (<30 minutes' duration) that is stable and not dehydrated
   - Owner is extremely worried

7. The case of the respiratory rat
   - Older rat, moderately obese, in respiratory distress
   - No upper respiratory congestion, other 5 rats in household asymptomatic

8. The case of the gasping galah
   - Open mouth breathing with neck extended, marked biphasic respiratory effort

9. The case of the painful pig
   - Intact female guinea pig that is straining to urinate
   - Patient's bladder is small and expressible, but she vocalizes on palpation and is straining
   - Urine is pink tinged

10. The case of the hemorrhagic hedgehog
    - Hedgehog comes in for second opinion, veterinarian 1 month ago saw it for a "urinary tract infection" and started antibiotics, bleeding went away but now it is back
    - Patient is producing massive amounts of blood clots but running around and acting normal

11. The case of the dying dragon
    - Bearded dragon that hasn't eaten in 1 year
    - Patient is emaciated and near death

Getting the Most Out of Our Journals: Finding the Evidence in a Peer-Reviewed Paper

Mark A. Mitchell, DVM, MS, PhD, DipECZM (Herpetology)

Session #2701

Affiliation: From Louisiana State University School of Veterinary Medicine, Skip Bertman Drive, Baton Rouge, LA 70803, USA.

Abstract: We all have a desire to use evidence-based medicine to manage the needs of our patients. Unfortunately, we are in a field where objective, evidence-based data are lacking. To counter this, it is important for all professionals in our field to contribute new knowledge. One of the reasons more research is not done has been attributed to a lack of understanding of study design and statistics. The author believes that these types of obstacles can be overcome if veterinarians approach these challenges as they do veterinary medicine. Veterinarians are trained to use hypothetico-deductive reasoning and decision tree analysis when approaching their cases. Patterns evolve out of these methods which help grow our confidence as veterinary scientists. Similar patterns exist with study design and statistics. By following these similar thought processes, it is possible for veterinarians to gain the necessary confidence they need to perform these important tasks and contribute to our body of evidence-based medicine. In addition, knowledge of these methods can also improve the veterinary scientist's ability in interpret the peer-reviewed literature. As study flaws are common, understanding the limitations of a peer-reviewed paper can minimize the likelihood of incorporating this flawed information into our knowledge base. The goal of this masterclass is to provide attendees with an introduction to the methods used by the author to perform these important tasks. By learning these patterns, attendees can increase their confidence and understanding of study design and statistics.

Introduction

Study design

Success with research starts with the design of a study. Many of the failures clinicians experience can be directly related to a lack of planning during the original design and development of a study. The types of studies performed in veterinary medicine are varied and may include basic bench top research to clinical trials. With such a broad range of topics and hypotheses to be tested, it should not be unexpected that the types of studies required to answer these scientific questions can vary. Developing a sound understanding of the different types of studies available and when/how to use them to answer specific research questions is essential to the success of an individual(s) planning a research project.

There are four basic study designs that will be discussed: cross-sectional studies, case-control studies, cohort studies, and experimental studies. To decide which study design is best, first consider the types of data to be collected and the specific hypotheses being proposed. Other considerations when planning a study include the feasibility of collecting the data, any temporal components associated with the study, and the resources (both financial and personnel) available. If you can answer these questions in the early stages of planning a study, it will help ultimately determine which study design will be best fit your needs.

Cross-sectional studies are an excellent first step for attempting to ascertain the occurrence of some outcome (e.g., disease process) and any associated independent variables in a population. This type of study is commonly referred to as a prevalence study. Cross-sectional studies evaluate a study population at a given point ("snapshot") in time. These types of studies are routinely pursued in veterinary clinical medicine because the data can be collected from patients as they present to determine the prevalence of some occurrence. These types of studies are well suited for those research questions attempting to identify how common or uncommon a disease might be or whether there are any inherent risk factors associated with a specific disease. A deficiency of these types of studies is that they do not provide information relative to the temporal nature of a disease in a case/population. For example, if you are interested in estimating the prevalence of *Ophidiomyces* in snakes by evaluating PCR results, you will be able to determine which snakes are positive or negative and which, if any, risk factors (e.g., species, sex, activity) predispose them to an outcome of being positive or negative; however, you will be unable to determine when the snakes were exposed to the fungus or for how long. Of course once you have completed a cross-sectional study and determined such a disease was prevalent, follow up studies (e.g., cohort studies) could be designed to further evaluate a temporal question.

Case-control studies are commonly done in veterinary clinical medicine because they are often designed to be retrospective in nature and can utilize historic (already available) information. These types of studies can provide a great deal of information regarding risk factors for a disease or outcome without needing a large amount of research dollars. With case-control studies the disease outcome is known and the relative effect different risk factors have on that outcome can be assessed by matching the patients to conspecifics that are presented for different outcomes. This type of study is well suited for preliminary investigations (pilot studies) that hope to collect sufficient results to pursue research grants for more developed studies or as a stand-alone study that can contribute to evidence based medicine.

Cohort studies are not commonly done in veterinary clinical medicine because of their longitudinal design and cost. Whereas a case-control study evaluates a known outcome and seeks to determine the effect of specific risk factors on that outcome, cohort studies seek to longitudinally evaluate the potential outcome(s) that occur as a result of various risk factors. This type of study should not be used for rare disease events, as it might not be possible to measure a specific outcome within a reasonable amount of time. These types of studies have contributed significantly to human medicine, and would likely do the same in veterinary medicine if the resources were available to pursue them.

Experimental or intervention studies are one of the most common types of studies pursued in veterinary clinical medicine. This type of study, when done properly, places the control of the research in the hands of the investigators. Experimental studies can range from being as simple as evaluating a simple outcome while controlling a single independent variable (e.g., one group receiving a treatment and another a placebo) to more complex cross-over studies evaluating multiple independent variables and outcomes.

Regardless of the study design being used, it is imperative to consider the sample size required for the study, the methods used for the randomization of the study subjects, and the potential for introducing and controlling different types of bias that could have negative effects on the results. Something that is expected for most study proposals through granting agencies and institutional animal care and use committees is the pre-determination of the sample size required to answer specific hypotheses. In the past, investigators might get by using examples of previous studies to estimate their sample size for a study; however, this is not considered sufficient and pre-determining sample size using a priori information specific to the sample population and the data collected are now required. Being able to have a sound estimate for a sample size for a study can go a long way in limiting problems later (e.g., introduction of a type II error). Random allocation of subjects into test groups should be done using appropriate methods, such as random number generators. Software to perform this is readily available, so it is now possible to be thorough and random when implementing a study. Finally, bias is a major contributor to study failure. Pre-evaluating a study for these types of error can help minimize the likelihood of them occurring, which will increase the overall power in the results.

There is always a great deal of work required to complete a research study. To ensure that the results of the study will have value and contribute to evidence-based medicine, it is important to design a study that will answer the specific hypotheses being proposed. Taking time to consider the potential pitfalls that might occur in a study during the initial planning stages will greatly improve an investigator's likelihood of success.

Parametric versus Non-Parametric Statistical Tests

The first step in selecting statistical tests is to determine which group of statistics to use. For simplicity, we will divide the statistics into two categories: parametric and non-parametric statistics. Certain assumptions must be met in the study design to properly perform and utilize a statistical test. Two of the most common assumptions, which are needed in both parametric and non-parametric tests are: (1) data collected for analysis should be obtained from study subjects randomly selected from the population represented in the study and (2) the individuals and their data should be independent of one another. These represent basic tenets to follow regardless of the statistics used. Note other assumptions must be met for each specific statistical test, the experimental design should reflect these assumptions. A major difference between the two types of statistics is related to the distribution of the data; parametric statistics should only be used on data that meets the assumption of normality, while non-parametric statistics do not have this restriction. Thus, if we simplify this to the types of data, it is obvious that categorical and ordinal data must always be analyzed using non-parametric statistics and that continuous data that doesn't meet the assumption of normality must also be analyzed using non-parametric statistics. The only data type that we would use parametric statistics for are those that meet the assumption of normality. Now, it is important to point out that continuous data that doesn't meet the assumption of normality can be transformed (e.g., log transformed) to potentially normalize the data. If this transformation is successful, then the more robust parametric statistics could be used to analyze the data; however, it would still be important to report the data by their median and percentiles as the transformed data would not be clinically relevant.

Statistical software

Statistical software is available that has simplified the process for analyzing data. These software programs come in a variety of packages, from those that require programming to those that don't require programming. The programs that require programming tend to be more robust in what they can accomplish. Examples of these types of programs include SAS (SAS Institute Inc., Cary, NC USA) and R (https://www.r-project.org/). R is free. MedCalc and SPSS (current version 24.0; IBM Statistics, Armonk, NY USA) do not require programming, although programming can be done with SPSS when more involved statistical methods are required. The author uses a combination of these programs. For those running very basic statistics, such as chi square tests and Fisher exact tests, free calculators are available on the Internet. The Centers for Disease Control and Prevention also have a free software program, EpiInfo (https://www.cdc.gov/epiinfo/index.html). Regardless of the software being used, it is important to understand when to select a test to ensure the analysis is correct, as these software programs will generate results for whatever test is selected, even if it is the wrong test.

Summary

The field of zoological medicine requires evidence-based knowledge to expand our understanding of these magnificent animals. Evidence-based knowledge requires hypothesis generated research that is validated using appropriate study design and statistical analysis. Every member of our profession can and should feel an obligation to contribute the generation of evidence-based science. For it is only through this new knowledge that we can achieve our potential as a true specialty.

Section 36

Hybrid: Emergency and Critical Care

In Honor of Marla Lichtenberger, DVM, DACVECC

Anna-Marie Ruoff, DVM;
Julianne Vickstrom, DVM
Session Chairs

Wildlife Rehabilitation: Rapid Diagnosis!

Renée Schott, DVM, CWR

Session #2519

Affiliation: From Wildlife Rehabilitation Center of Minnesota, 2530 Dale St. N, Roseville, MN 55113, USA.

Abstract: This fun and interactive class utilizes real cases and many pictures to review some of the most common presentations of wildlife and the diagnoses associated with those cases. Be prepared to participate, laugh, and maybe even learn something! Species covered include songbirds, waterfowl, waterbirds, turtles and more! Types of cases span those with simple diagnoses and quick turnarounds to complicated cases with involved treatment plans.

Recommended Reading

1. Hernandez SM, Barron H, Miller E, et al, eds. *Medical Management of Wildlife Species.* Hoboken, NJ: Wiley-Blackwell; 2019. doi:10.1002/9781119036708.

2. Mullineaux E, Keeble E, eds. *BSAVA Manual of Wildlife Casualties.* 2nd Ed. Gloucester, England: British Small Animal Veterinary Association; 2016.

3. Friend M, Franson JC, eds. *Field Manual of Wildlife Diseases: General Field Procedures and Diseases of Birds.* Madison, WI: US Geological Survey; 1999.

4. Ruth I. *Wildlife Care Basics for Veterinary Hospitals: Until the Rehabilitator Arrives.* Washington, DC: Humane Society of the United States; 2012.

5. Gage L. *Hand-Rearing Wild and Domestic Mammals.* Ames, IA: Blackwell Publishing; 2002.

6. Miller E, ed. *Minimum Standards for Wildlife Rehabilitation.* 4th Ed. St. Cloud, MN: National Wildlife Rehabilitators Association and International Wildlife Rehabilitation Council; 2012.

7. Stocker L. *Practical Wildlife Care.* 2nd Ed. Oxford, England: Blackwell Publishing; 2005. doi:10.1638/1042-7260(2000)031[0586:br]2.0.co;2.

8. Meredith A, Keeble E. *Self-Assessment Colour Review: Wildlife Medicine and Rehabilitation.* London, England: Manson Publishing; 2011. doi:10.1201/b15226.

Basic Wildlife Rehabilitation Triage

Renée Schott, DVM, CWR

Session #2518

Affiliation: From The Wildlife Rehabilitation Center of Minnesota, 2530 Dale St. N, Roseville, MN 55113, USA.

Abstract: It can be difficult for a practitioner to know which wild animals will survive the rehabilitation process to release and which will not. This master class uses data from The Wildlife Rehabilitation Center of Minnesota, one of the largest wildlife rehabilitation centers in the country, to help predict which patients will make it to release and which will not. The goal of this class is to help those in private practice know when to euthanize on admission and when to treat and transfer to a wildlife rehabilitator.

Adapted from Schott R. Wildlife rehabilitation: Until the rehabilitator arrives. Proc ExoticsCon. 2020.

Introduction

Private practitioners are often regarded as experts across all taxa, regardless of the focus of their practice. Clients and the general public will occasionally call with questions about injured or orphaned wildlife, or may simply show up in the lobby with the animal. It behooves the clinic to be prepared to deal with these scenarios so the finders have resources and options. It is important that clinics are prepared to handle and refer these calls appropriately for the sake of efficiency, the welfare of the wild animal, and the safety of the patients in the hospital. In addition, it is important for private practitioners to be armed with basic triage and euthanasia skills to humanely end the suffering of wildlife that cannot recover.

The Wildlife Rehabilitation Center of Minnesota (WRCMN) admits > 19,000 sick and injured wild animals each year. Asterisks (*) below indicate older data from WRCMN and updated data from WRCMN is used during the live class to help others make informed decisions. It is difficult to compare data from one rehabilitation center to another due to differences in location, resources, capacity, species, diseases, etc. For more details about how the data was compiled, please contact renee@wrcmn.org.

Preparation

Before wildlife is in the lobby or calls start coming in, clinics should prepare by following the steps below.

Find a wildlife rehabilitator or facility to work with

It is illegal to possess native wildlife without the proper permits. Veterinarians must be working with a permitted rehabilitator in order to possess wildlife they are treating. In addition, most states require rehabilitators to work with a licensed veterinarian to keep their permits active. Rehabilitators are familiar with natural history and veterinarians know veterinary medicine; working together ensures the best outcome for the animal. In addition, rehabilitators can determine when an animal needs to be rescued and when it should be left alone. Calls are best referred to rehabilitators or facilities set up to field these questions, as questions are complicated, and answers aren't always straight forward.

Compile resources

Where can you refer phone calls or injured patients to? Do you know the addresses or hours of facilities in the area? Who is a contact at your state natural resources or fish and game department and at US Fish and Wildlife? Making a resources sheet and SOPs for your reception will limit miscommunications.

Know the laws

Typically, each state regulates who can rehabilitate wildlife and how. Federal regulations exist for migratory birds as well. While it is usually legal for veterinarians to treat and/or euthanize wildlife, they must transfer all migratory birds to a licensed wildlife rehabilitator within 24 hours of stabilization[1] and they cannot possess other species for more than 24-48 hours unless they are already working with a permitted rehabilitator. There are often additional guidelines and regulations surrounding threated or endangered species. Keep in mind, species that have a hunting season in your area must be treated as food animals; withdrawal times and the list of prohibited medications apply.[2] More details about legal responsibilities and restrictions can be found in Duerr and Whittington, 2017.[3]

Euthanasia

For a wild animal to allow capture, it is usually very debilitated. Release proportions vary according to species, age and injuries, but range from 24%-85%.[4-7] Euthanasia is common, necessary and humane. In most areas, the number of animals that need rehabilitation are too numerous for the available resources. When considering the below cases, it is important to consider these resources, the anticipated number/species/age of animals needing rehabilitation, rehabilitator/facility limits, and welfare. The ability of the rehabilitator, facility, or clinic to accept animals for rehabilitation will be dependent on the above factors. If resources (time, space, money, people) are low, the overall ability of the caretakers will diminish which will lower welfare for everyone. Euthanizing the least likely to survive and basing the number of animals in care on resources available, will allow those with the best prognosis to receive the best care and ultimately the best chance at release. Because stretching resources results in a lower standard of care for all, euthanizing more animals on admission in a strategic manner results in an increase in the overall number of animals being released (*).

What types of orthopedic conditions should be euthanized?

Prognoses and treatment options are taxa specific.[8] Consider your ability, the ability of the wildlife rehabilitator, and the species. Below are generalizations that apply to most taxa unless specified otherwise.

Necrotic or old, compound fractures: Compound fractures are inherently more prone to healing complications owing to the increased likelihood of contamination and/or infection. Necrotic bone must be removed and in general, most species need to have almost perfect fracture alignment with < 10% loss of long bone length. Necrotic bone is almost always accompanied by necrotic soft tissue. Many wild animals (especially birds) have very little soft tissue over certain bones and the loss of that soft tissue means the bones underneath will not heal or will have complications.[9] Older fractures are also inherently prone to more complications; if the bone has stopped healing, then all bone ends must be freshened to be stimulated to heal again. In addition, muscles, tendons and ligaments often have contracted, making reduction difficult. Not only do the bones need to heal perfectly, but all joints need near perfect range of motion. The definition of "old fracture" varies with species, age and type of fracture. For instance, a two-day old compound, comminuted fracture in a nestling songbird may be considered old, whereas a 7-day old closed, simple, midshaft fracture in a fox may be fresh enough to fix.

Traumatic joint injuries: In the author's experience, wildlife that experience traumatic joint injuries, begin trying to reduce the abnormal movement in the joint by producing fibrous tissue around the joint. This reduces range of motion chronically which leaves the animal non-releasable. In the author's experience, luxations and sub-luxations (excluding coracoid luxations off the keel, which usually heal well) will almost always heal with reduced range of motion and should be euthanized. In addition, traumatic joint luxations frequently will cause osteoarthritis.[9,10]

Successful surgical management of elbow luxations in raptors[11] and a woodchuck[12] have been reported; however, in both papers the possibility of osteoarthritis was not discussed. Unsuccessful treatment of luxations in wildlife are also reported.[12,13] Articular fractures need perfect alignment and rigid fixation to prevent secondary degenerative joint disease.[14] Wildlife develop reduced range of motion and signs of osteoarthritis within weeks of sustaining an articular fracture, no matter how rigid the fixation and perfect the anatomic alignment. WRCMN euthanizes all traumatic joint injuries that result in intra-articular fractures or luxations/subluxations of joints (excluding coracoid luxations off the keel) because of past experiences (*).

Shoulder fractures in aerial insectivores: in North America, swallow species (including but not limited to Barn Swallows *Hirundo rustica* and Cliff Swallows *Petrochelidon pyrrhonota*), swift species (including but not limited to Chimney Swift *Chaetura pelagica* and Vaux's Swift *Chaetura vauxi*), nightjar species (including but not limited to Common Nighthawk *Chordeiles minor* and Eastern Whip-poor-will *Caprimulgus vociferus*) have very high metabolic demands; Vaux's swifts may need to eat >5000 arthropods each day![15] Flight must be perfect in order to accomplish this. A bird with even the slightest flight deficit may die of starvation after release. In addition, most of these species migrate to Central or South America for the winter which also requires sustained flight. To adequately assess pre-release flight ability, these species need a 30' (swallows/swifts) and 50' (nightjars) flight cage. Any deficit should be heavily scrutinized. In the author's and other's experiences (Veronica Bowers, Native Avian Care, personal communication, 2017) < 1 % of shoulder fractures in these species will recover to a releasable state. Several species do not eat in captivity and must either be trained to accept handfeeding, gavage or force fed. All are high stress species. Feather condition must be pristine. Husbandry for these species is complicated and best not undertaken by a novice.

All aerial insectivores with any shoulder girdle (coracoid, scapula, clavicle) fractures are now euthanized at WRCMN due to past experiences, other's experiences and the intense husbandry care these species require (*).

What type of traumatic eye injuries should be euthanized?

In general, any condition that will result in permanent vision loss should be considered for euthanasia, even if only in one eye. Prey animals will be left unable to watch for predators, predators may have a difficult time catching enough prey (no live prey testing in captivity is equitable to catching prey in the wild). This is a controversial topic and some wildlife rehabilitation veterinarians and rehabilitators have differing opinions. This author does not release one-eyed animals, with an exception being some aquatic turtle cases. Traumatic eye injuries that result in vision loss include luxated lens (posteriorly and anteriorly luxated), ruptured globes, complete retinal detachment with incomplete reattachment, and any other condition that results in permanent vision loss.

What types of other issues should be euthanized?

Chronic pain: Wild animals have evolved to hide signs of pain and to self-repair injuries as much as possible. If they did the opposite, they would likely become prey. Therefore, any condition which has the potential to result in chronic pain in a human, should be assumed to have the potential to result in a chronic pain for the animal, should be grounds for euthanasia of the wild animal for humane reasons. Wild animals cannot receive chronic pain medications or other palliative care in the wild.

Spinal trauma with no deep pain: Most rehabilitators and veterinarians recognize spinal trauma via pelvic limb paresis or paralysis. Because of the anatomy of the spinal cord, loss of deep pain perception indicates a grave prognosis for return to function. In addition, neuropathic pain has been reported in paralyzed companion animals.[16] Therefore, animals without deep pain should be humanely euthanized. To test for deep pain perception, use a toe on the limb that is paralyzed/paretic. Pinch the toe; start with a superficial pinch and then slowly pinch harder. Watch for head movement, vocalization or other indication that the pain sensation made it up the spinal cord and to the brain. Ignore all movement in the limb that is being pinched as the withdrawal reflex may be intact or hyper reflexive, depending on where the lesion is. One cannot discern between the withdrawal reflex and voluntary movement when pinching a toe, and wild animals can have very strong reflexes, so it is imperative to ignore movement in the leg and watch the head of the animal. In addition to spinal trauma, brachial plexus avulsion occasionally occurs. If deep pain is not present in one forelimb of a unilaterally forelimb paralyzed animal, brachial plexus avulsion should be considered. Brachial plexus avulsion will not heal to a releasable state and these animals should be euthanized. Animals with no deep pain in a limb should be euthanized as they will not recover to a releasable state.

Seizing rabies vector species: Rabies vector species (raccoon, coyote, fox, bat, skunk) with abnormal neurologic signs should be considered for euthanasia due to public health concerns. It is important to know your rabies vector species and rabies virus variants in your area and what your department of health requires you to do; they may want the animal tested for rabies or to have the contact of the finder to talk about exposure. Generally, a seizing animal bears a poor prognosis, regardless of species.

Distemper virus infection suspects: Distemper suspect species (i.e., raccoon, coyote, fox) with abnormal neurologic or severe respiratory clinical signs, with or without leukopenia should be considered for euthanasia. Consider the infectious nature of this pathogen and the domestic animals in your clinic which are susceptible. In addition, wild animals have a high probability of permanent neurologic issues even if they survive, which is rare.

Only 6.7% (25) of the 373 adult raccoons admitted to WRC from 2011-2017 were able to be released (*). Treatment was attempted with 68 and 25 (36.8%) of treated were released. While not all of these cases were distemper suspects, adult raccoons are difficult for the public to capture and transport; consider how debilitated this species must be to present for rehabilitation and this bears out in prognosis.

Infants: The younger the animal is, the lower it's chance of survival when being raised in captivity by humans. Reviewing data from previous years can help facilities find which variables are best correlated with survival to release.

WRCMN data (*):

- Eastern cottontail rabbits (*Sylvilagus floridanus*): WRCMN received 3133 young cottontails in 2017 alone. Historically, under 65 grams, healthy Eastern cottontail rabbits have < 30% chance of release. Under 75 grams, unhealthy Eastern cottontail rabbits have a < 30% chance of release. Definition of healthy: mild-moderate dehydration, thin (not emaciated), BAR, no injuries.

- Eastern grey squirrels (*Sciurus carolinensis*): WRCMN received 1285 young grey squirrels in 2017 alone. Historically, healthy grey squirrels under 40 grams have < 30% chance of release.

- Virginia opossums (*Didelphis virginiana*): WRCMN received 119 young Virginia opossums in 2017. Opossums 20-40g have a 50% chance of survival.

- Hatchling altricial birds (just out of shell) are euthanized as they have 0% chance of release at WRCMN.

Adult Eastern cottontails and Eastern grey squirrels with abnormal brain neurologic clinical signs
Adult Eastern cottontails and Eastern grey squirrels commonly present to WRCMN with abnormal brain neurologic clinical signs (e.g., head tilt, circling, head turn, ataxia). Necropsy results most commonly conclude the etiology to be *Baylisascaris procyonis* (or a nematode tract through the brain) or head trauma.

WRCMN received 233 adult cottontails with abnormal central nervous system signs in 2017 (*). From 2011-2017, 1727 total cottontail adults were admitted. Of these 641 had abnormal brain neurologic clinical signs. Of the 641, 3.3% were eventually released. Of the ones that were treated, 8% were released. This is compared to all other adult cottontails which had an overall release proportion of 5.5% and a treated released proportion of 19%. This author recommends euthanasia of all adult cottontails with abnormal brain neurologic clinical signs due to the poor release proportion.

WRCMN received 94 adult grey squirrels with abnormal central nervous system signs in 2017 (*). From 2011-10.22.2017, 1295 total grey squirrel adults were admitted. Of these 414 had abnormal brain neurologic clinical signs (such as head tilt, circling, head turn, ataxia, etc). Of the 414, 8.7% were eventually released. Of the ones that were treated, 16.1% were released. This is compared to all other adult grey squirrels which had an overall release proportion of 10% and a treated released proportion of 29.1%. This author recommends heavily scrutinizing any adult Eastern grey squirrel with abnormal brain neurologic signs before choosing to treat.

Conditions to Not Euthanize (if you have the resources)

Mycoplasma gallisepticum conjunctivitis in house finches

Mycoplasma gallisepticum (MG) originally spread from the production poultry industry to house finches (*Haemorhous mexicanus*) in the 1990s[17] and is now endemic in the house finch population.[18] When it first appeared, the poultry industry and biologist were afraid of MG recovered house finches becoming latent carriers and some antibiotics being ineffective;[19,20] however, it is very hard for wild house finches to actually transmit *M. gallisepticum* to domestic poultry[21,22] and there is a treatment, that when given appropriately can eliminate the carrier state.[23] Euthanizing one or even several hundred house finches because they have this disease will not affect the population or disease status. WRCMN treats and releases house finches with presumed MG conjunctivitis according to Mashima et al 1997.[23]

Severe mange in red fox

Juvenile red fox (*Vulpes vulpes)* occasionally are admitted in the fall with severe mange, presumably after having been forced out of maternal territory. While some of these fox are too sick or emaciated with secondary conditions to survive, many can be treated and released.[24] These patients require intensive critical care as they are usually severely emaciated, chronically anemic, and have a myriad of other health concerns that will develop during treatment (i.e., refeeding syndrome).

Animal missing digits

Certain species can survive well in the wild with missing digits; however, the location and number of missing digits that allow successful release are species dependent.[8]

How to Euthanize Various Wildlife Species

The techniques described below are used most frequently at WRCMN. They take into consideration the wild animal species, age, and physiology of the animal being euthanized as well as the safety of the handler; priority is given to decreasing apparent animal stress. There are other technique options. In the author's experience, the methods below have proven reliable and cause the least amount of stress to the animal.

Euthanasia Techniques

Isoflurane (or sevoflurane) via face mask

Restrain the patient and place head in mask through diaphragm or hook up vaporizer to an induction box with the animal inside. Start at maximal percentage until patient is fully anesthetized. Then administer appropriate amount of pentobarbital IV or IC.

Isoflurane box

- Liquid isoflurane is poured into a clear Rubbermaid-type container with a lid. Either do this in a vent hood or ensure container is on the ground/floor and you remain above to prevent accidental inhalation of isoflurane, which is heavier than room air. Ideally, a respirator would be worn to avoid inhalation of all fumes when not in the hood.
- If the animal is fractious, aggressive or high stress, ideally the container would be big enough for a live trap or kennel. Having several sizes of containers would be ideal for different sized animals.
- Use 1 tbsp liquid isoflurane for every 4 gallons container volume.
- The box should be mostly sealed (but not completely sealed as the gas expands).
- After isoflurane is poured into the container, the animal (or animal in a live trap) can be transferred to the container and the lid replaced.
- Monitor the patient to ensure it quickly lose consciousness.
- Leave patient in the tub for at least 10 minutes to ensure over dose of isoflurane has caused death. Overdose of isoflurane is an approved humane euthanasia method according to the AVMA.[25]
  - Pinkie mammals may need longer depending on their body temperature and age.
- Rock tub before opening to ensure no movement.
- Open tub outside (tub on ground, ensure you are above tub or are wearing a respirator) or in a vent hood.
- Use a stethoscope to ensure heartbeat of the patient has stopped.
- Leave tub outside or in vent hood for at least 10 minutes.

Table 1. Euthanasia choices for wildlife.[a]

Species/group	Age	First choice	Second choice	Notes
Eastern cottontail rabbit	Adult (250 g-1.5 kg)	0.2 ml (100 mg/ml) ketamine; 0.4 ml (5 mg/ml) midazolam; 0.2 ml (10 mg/ml) butorphanol → mix→ give IM → wait 15 min. If still alive, put into isoflurane box[b] until deceased or mask at 5% isoflurane until anesthetized then IC sodium pentobarbital (390 mg/ml) 0.5 ml	2 ml (100 mg/ml) alfaxalone → give IM → wait 15 min If still alive, put into isoflurane box[b] until deceased or mask at 5% isoflurane until anesthetized then IC sodium pentobarbital (390 mg/ml) 0.5 ml	Prolonged struggling and breath holding when placed in isoflurane box without pre-medications and even more prolonged in the isoflurane face mask/induction box
Other mammal species	Neonate, just weaned	Isoflurane box[b] until deceased. They dislike the strong smell and will wipe their faces but are anesthetized within 10 seconds		Young mammals' brains are very tolerant to hypoxia and they hold their breath and struggle in an isoflurane mask at 5% for several minutes
Breath-holding avian species (i.e., diving species, waterfowl, waterbirds)	Adults	3 mg/kg midazolam, 5 mg/kg butorphanol, 5 mg/kg ketamine IM → wait 15 min then isoflurane box[b] until deceased	Debilitated individuals are often too weak to hold their breath and do well in the isoflurane box[b]	
Avian species	Any (except breath-holding species)	Isoflurane box[b] until deceased		Birds have efficient gas exchange in their lungs; they become anesthetized in ~2 breaths
Raccoons/ similar sized canids	Adults	0.3 ml (0.5 mg/ml) dexdormitor; 0.3 ml (10 mg/ml) butorphanol→ mix→give IM; if seizing, start with 1mg/kg midazolam→ wait 15 min, if still alive, put into isoflurane box[b] until deceased or mask at 5% isoflurane until anesthetized then IC sodium pentobarbital (390 mg/ml) 3 ml	Debilitated individuals are often too weak to hold their breath and do well in the isoflurane box[b]	
Aquatic turtles	Any	15-25 mg/kg alfaxalone IM →wait 15 min → 10 mg/kg Propofol IV wait 15 min→ same volume (as propofol) pentobarbital IV wait 15 min → pith	Other anesthesia medications can be used but all must induce anesthesia and end in pithing	Chelonian brains are so tolerant to hypoxia[26-28] that reanimation can happen if using anesthetics/ pentobarbital alone

[a]The Wildlife Rehabilitation Center of Minnesota (WRCMN) admits > 19,000 sick and injured wild animals each year. Asterisks (*) are used to indicate WRCMN data.
[b]See isoflurane box euthanasia technique below.

Basic Wildlife Stabilization

During all of these steps, decreasing stress is the most important thing we can do for wildlife. Stress will kill wildlife and often a hands-off protocol is preferable over an invasive one.

First. Do visual triage. Is there an obvious injury that would render the animal non-releasable? If so, euthanize immediately. Immediately treat any breathing difficulties (place in oxygen chamber) or active hemorrhage (apply bandage to stop bleeding).

Second. Treat thermoregulatory deficits. If animal is hypothermic, warm it up. If animal is hyperthermic, cool it down.

Third. Leave the animal alone! Leave the animal in a quiet, dark room so it can de-stress and relax. This also gives the animal time to adjust internal body temperature.

Fourth. Give fluid therapy. Assume at least 5% dehydration in all animals admitted. Subcutaneous fluids are fast to administer and a great choice in almost all cases.

Fifth. Give pain medications. If there is any injury, appropriate pain management is paramount.

References

1. Code of Federal Regulations. 2018. Available at: https://www.ecfr.gov/cgi-bin/text-idx?SID=0e9fd773e2a078e4c54d159ce4a9be80&node=50:9.0.1.1.4.2.1.2&rgn=div8. Accessed June 1, 2022.

2. Schott R. Extra-label drug use in wildlife rehabilitation medicine. *Wildl Rehabil Bull.* 2017;35:(2)33-36.

3. Duerr R, Whittington JK. Legal responsibilities and restrictions on veterinarians working with wildlife. *Wildl Rehabil Bull.* 2017;35(1):25-37.

4. Kelly A, Bland M. Admissions, diagnoses, and outcomes for Eurasian sparrowhawks (*Accipiter nisus*) brought to a wildlife rehabilitation center in England. *J Raptor Res.* 2006;40(3):231-235. doi:10.3356/0892-1016(2006)40[231:ADAOFE]2.0.CO;2.

5. Molina-López RA, Casal J, Darwich L. Final disposition and quality auditing of the rehabilitation process in wild raptors admitted to a wildlife rehabilitation centre in Catalonia, Spain, during a twelve year period (1995-2007). *PLoS ONE.* 2013;8(4):1-8. doi:10.1371/journal.pone.0060242.

6. Barrows Z, Fix S. Raptors rehabilitated study in Iowa. *Wildl Res.* 1990;26(1):18-21.

7. Orós J, Montesdeoca N, Camacho M, et al. Causes of stranding and mortality, and final disposition of loggerhead sea turtles (*Caretta caretta*) admitted to a wildlife rehabilitation center in Gran Canaria Island, Spain (1998-2014): a long-term retrospective study. *PLoS ONE* 2016;11(2):1-14. doi:10.1371/journal.pone.0149398.

8. Duerr R, Purdin G, eds. *Topics in Wildlife Medicine: Orthopedics*. St Cloud, MN: National Wildlife Rehabilitators Association; 2017.

9. Johnson AL. Fundamentals of orthopedic surgery and fracture management. In: Fossum TW, Hedlund CS, Johnson AL, et al., eds. *Small Animal Surgery.* 3rd Ed. St. Louis, MO: Elsevier; 2007:930-982.

10. Denny HR, Butterworth S. Osteoarthritis. In: Denny JR, Butterworth S, eds. *A Guide to Canine and Feline Orthopaedic Surgery.* 4th Ed. Hoboken, NJ: Wiley-Blackwell; 2000:52-63.

11. Ackermann J, Redig PT. Surgical repair of elbow luxation in raptors. *J Avian Med Surg.* 1997;11(4): 247-254.

12. Chambers JN, Hanley CS, Hernandez-Divers S. Hip luxation in a woodchuck (*Marmota monax*): successful treatment by closed reduction and a modified Ehmer sling. *J Zoo Wildl Med.* 2004;35(4):569-571. doi:10.1638/04-012.

13. Bennett KR, Desmarchelier MR, Bailey TR. Surgical correction of bilateral patellar luxation in an american black bear cub (*Ursus americanus*). *J Zoo Wildl Med.* 2015;46(2):359-362. doi:10.1638/2013-0288R1.1.

14. Newton CD. Etiology, classification, and diagnosis of fractures. University of Pennsylvania. 2018. Available at: http://cal.vet.upenn.edu/projects/saortho/chapter_11/11mast.htm#b. Accessed November 4, 2021.

15. Bull EL, Beckwith RC. Diet and foraging behavior of Vaux's swifts in northeastern Oregon. *Condor.* 1993;95(4):1016-1123. doi:10.2307/1369437.

16. Granger N, Carwardine D. Acute spinal cord injury: tetraplegia and paraplegia in small animals. *Vet Clin North Am Small Anim Pract.* 2014;44(6)1131-1156. doi:10.1016/j.cvsm.2014.07.013.

17. Citti C, Blanchard A. Mycoplasmas and their host: emerging and re-emerging minimal pathogens. *Trends Microbiol.* 2013;21(4):196-203. doi:10.1016/j.tim.2013.01.003.

18. Dhondt AA, DeCoste JC, Ley DH, Hochachka WM. Diverse wild bird host range of *Mycoplasma gallisepticum* in eastern North America." *PLoS ONE.* 1014;9(7). doi:10.1371/journal.pone.0103553.

19. Wellehan JFX, Zens MS, Calsamiglia M, et al. Diagnosis and treatment of conjunctivitis in house finches associated with mycoplasmosis in Minnesota. *J Wildl Dis.* 2001;37(2):245-251. doi:10.7589/0090-3558-37.2.245.

20. Williams ES, Yuill T, Artois M, et al. Emerging infectious diseases in wildlife. *Rev Sci Tech.* 2002;21(1): 139-157. doi:10.20506/rst.21.1.1327.

21. Luttrell MP, Stallknecht DE, Kleven SH, et al. *Mycoplasma gallisepticum* in house finches (*Carpodacus mexicanus*) and other wild birds associated with poultry production facilities. *Avian Dis.* 2001;45(2):321. doi:10.2307/1592971.

22. Stallknecht DE, Luttrell MP, Fischer JR, Kleven SH. Potential for transmission of the finch strain of *Mycoplasma gallisepticum* between house finches and chickens. *Avian Diseases* 1988;45:352-358.

23. Mashima TY, Ley DH, Stoskopf MK, et al. Evaluation of treatment of conjunctivitis associated with *Mycoplasma gallisepticum* in house finches (*Carpodacus mexicanus*). *J Avian Med Surg.* 1997;11:20-24.

24. Newman TJ, Baker PJ, Harris S. Nutritional condition and survival of red foxes with sarcoptic mange. *Can J Zool.* 2002;80(1):154-161. doi:10.1139/z01-216.

25. American Veterinary Medication Association. *AVMA Guidelines for the Euthanasia of Animals: 2013 Edition*. Available at: https://www.purdue.edu/research/oevprp/regulatory-affairs/animal-research/docs/2013_AVMA_Guidelines.pdf. Accessed June 1, 2022.

26. Jackson DC, Ultsch GR. Physiology of hibernation under the ice by turtles and frogs. *J Exp Zool A: Ecol Genet Physiol.* 2010;313A(6):311-327. doi:10.1002/jez.603.

27. Larson J, Drew KL, Folkow LP, et al. No oxygen? No problem! Intrinsic brain tolerance to hypoxia in vertebrates. *J Exp Biol.* 2014;217:1024-1039. doi:10.1242/jeb.085381.

28. Ultsch GR. The ecology of overwintering among turtles: where turtles overwinter and its consequences. *Biol Rev Camb Philos Soc.* 2006;81:339-367. doi:10.1017/S1464793106007032.

Working with Exotic Pet Rescues and Shelters

Cathy A. Johnson-Delaney, DVM

Session #2678

Affiliation: From Washington Ferret Rescue & Shelter, 12522 128th Lane NE, Kirkland, WA 98034, USA.

Abstract: Bird and exotic animal rescues and shelters have need of veterinarians with experience not only with single pet birds and exotics but with animal colonies as well as general shelter expertise. Resources including facility, finances and personnel are often extremely limited. Preventive health program development as well as training of personnel, assistance with facility planning, development of standard operating procedures, regulatory issues interpretation and guidance, establishment of an animal database and records systems, and development of the organization's culture are part of what the exotic animal veterinarian can provide beyond just medical and surgical care. The veterinarian needs to establish clear guidelines for the amount of involvement that can be provided as well as fees for any services, pharmaceuticals, imaging and other equipment usage.

Introduction

Most exotic pet veterinarians will be asked to help with an exotic pet rescue and shelter, as more of these organizations are being formed. And unlike government or major animal charity organization that have a full or part-time veterinarian on staff, many of these are small and run by individuals, and cannot afford full or part-time staffing but turn to a veterinarian on an as needed basis. This master class is primarily discussing the small non-dog, non-cat exotic animal rescues and shelters that do not regularly employ a shelter veterinarian, but rather try to work with a local exotics veterinarian to provide for their facility. The veterinarian fills a unique role in assuring that these species receive appropriate diet, husbandry and medical care. How involved in the actual organization the veterinarian is becomes a labor of love, as most time spent is gratis, although supplies, facility/equipment usage, technician time, diagnostics, and pharmaceuticals used may be compensated often at a deep discount or as negotiated fees. Ferrets, rabbits, small rodents, hedgehogs, sugar gliders, reptiles and birds are increasingly being relinquished to shelters as owners can no longer care for them. Because these species are usually not accepted at municipal and major humane facilities designed for dogs and cats, many exotic pet shelters are run by individuals. Without governmental funding, most of these private shelters depend on donations, adoption fees, and volunteer participation to care for these animals. The veterinarian can be instrumental in educating and training the volunteers about the proper care of these species.

Shelter medicine its own field of veterinary medicine, although primarily it is geared to dog/cat or equine facilities, with veterinarians usually paid for services.[1-4] These are usually linked to municipal or government animal control agencies or private humane facilities operated by or associated with national or international humane organizations. Veterinarians working with shelters have to change from primarily working with individual animals and owners to a more herd health mentality as far as realistic management from a space, person-power, budget, and time consideration.[1-4] Working with shelters that cater to the non-dog, non-cat abandoned animal face even more obstacles as husbandry, dietary, medical and even zoonotic disease issues of the different species are extremely variable. They also rarely receive support from community or governmental sources. Guidelines are beginning to be included in some of the major shelter medicine texts, as well as there now exists the Association of Shelter Veterinarians as well as an American Board of Veterinary Practitioners (ABVP) Certification in

Shelter Medicine.[5-9] There also is a program for Shelter Medicine at UC Davis.[10] For purposes of this discussion, sanctuaries are facilities that take in unwanted or neglected animals to care for them for the life of the animal. Shelters operate with the goal of re-homing the animals.

An Emerging Necessity: Exotic Pet Shelter Medicine

"The good of the many outweighs the good of the few, or the one."[11] While many volunteers and animal lovers may find this quote contrary to their beliefs and reasons for establishing or assisting a shelter, the reality is that there are always far more animals being abandoned and needing care than shelters can maintain for the life of the animal. Most avian and exotic pet shelters would consider themselves no-kill shelters (i.e., that the animal will be maintained until it is adopted or fostered). No-kill shelters must still have in place a policy to humanely euthanatize an animal presented in grave condition or one that is very ill or in pain that cannot be readily treated given the funding, staffing, and resources of the shelter. Shelters must establish a policy of how much of their budget can go to treating any particular animal. This is one of the aspects of shelter medicine that is the most difficult for volunteer caretakers: the decision of when to euthanatize an animal, who makes that decision, and the accepted humane method as well as the policy for disposal of the body. This policy needs to include not only the animals surrendered in grave condition, but those housed at the shelter as well as those in foster care homes. All policies must be in writing, and training sessions should be held on a regular basis. Shelters should have legal documentation of surrender of an animal to their care. Table 1 is a sample surrender form in current usage. Forms utilized by shelters for surrender, fostering and adoption should be reviewed by an attorney.

Shelters frequently begin in someone's home who takes in abandoned pets. While their initial reason to do so is compassion and care, without planning for increased numbers of pets within their home, they are easily and quickly overwhelmed. Most people work outside the home, and additional exotic animals require extra time as well as financial commitment. Individuals soon find they need help maintaining the animals, hence they recruit friends or other owners of the same exotic animal species to help. Individuals also soon find out they are tied down with so many animals, as word usually gets out that they will take that species, and more unwanted animals are surrendered. Those individuals who enlist the help of others to care for these animals usually reach the point where they realize they need to form some sort of organization to continue, particularly if they realize they cannot keep all the pets themselves. They need to develop a method to adopt them to appropriate homes. Individuals who just continue to collect animals in the guise of a shelter but do not enlist the help of others and begin forming an organization and network are hoarders. Hoarders or collectors are usually well-meaning but rarely can house, feed, and provide the needed medical care for these animals. An individual identified as a hoarder frequently comes to the attention of animal control agencies, public health agencies and even mental health agencies. Animals confiscated then need to be rehabilitated and placed.

Sanctuaries and shelters that tend to be longer lived are those that are located in a separate facility from any one volunteer or founder's home. Location of a facility must comply with zoning that will allow for an animal shelter. Most residential neighborhoods particularly within city limits have restrictions on the number of animals that can be maintained, although many do not specify caged pets. Shelters that organize and become 501(c) non-profit organizations also tend to be more stable, and less likely to get into the "hoarding" mode, although policies need to be in place to inspect foster animal placement homes, and to keep foster animals medically checked. Sanctuaries and shelters need to set a limit for the number of animals they can afford, have space to house, feed, and clean, in addition to being able to provide for veterinary care. Volunteers frequently take home more than they can really care for too, so setting up mandatory training programs for fosterers as well as a schedule of medical evaluations of the animals should be incorporated into the policies of the shelter. All policies and training programs should be reviewed annually, and training sessions held regularly.

Table 1. Washington Ferret Rescue & Shelter Surrender Agreement Form.

I/we____have made the decision to surrender the following named ferret(s) and have chosen to enlist the adoption services of the Washington Ferret Rescue & Shelter, hereafter designated as WFRS, that will find new homes as appropriate.

Name of ferret:________________________________ Color______________

Male/Female Age__________Neutered: Y/N CD Vaccination Date: ________________

Name of ferret:________________________________ Color______________

Male/Female Age__________Neutered: Y/N CD Vaccination Date: ________________

Name of ferret:________________________________ Color______________

Male/Female Age__________Neutered: Y/N CD Vaccination Date: ________________

Name of ferret:________________________________ Color______________

Male/Female Age__________Neutered: Y/N CD Vaccination Date: ________________

It is understood that the ferret(s) are being surrendered for their continued welfare and that the WFRS accepts said ferret(s) in an AS IS condition. It is further understood, therefore, that the WFRS assumes full legal ownership, thereby full responsibility, for any current or future illness while in the custody of the WFRS. It is also understood that items surrendered with the ferret(s) become the legal property of the WFRS, and as such will not be refunded nor returned.

Total confidentiality will be observed unless the surrendering party wishes to communicate with the *adopting* party, in which case, *this request must be put in writing* for the adopting party's acknowledgement.

The WFRS reserves the right of refusal in placing any ferret so surrendered if in their judgment certain prerequisites of ownership are not met. No ferret will be placed in any city, county, or state that restricts or bans ownership of ferrets specifically or by inclusion, nor shall any ferret be knowingly placed in a rental/condominium that excludes pet ownership, nor shall any ferret be knowingly placed in a home or institution for the purpose of ritual, rite, or medical experimentation.

The WFRS guarantees to the surrendering party that their pets will receive the utmost care and consideration in the placement with a new owner and that future contact with the same will be done in the interest of continued health and welfare concerns.

The WFRS operates this Shelter as a service to ferrets and as such, is not a business. It is a non-profit 501(c)(3) corporation.

Signed______________ Date________________
Name (Printed)____
Address____
Phone______________ Email____________________________

I would like to make a donation of $__________, to help cover the food and medical costs of the ferret(s) I am surrendering to the WFRS. Donations may be tax-deductible where applicable.

Received by WFRS staff member: ______________ Date ______________

The veterinarian's role

The shelters that organize will elect a board, and establish guidelines to govern acceptance of animals, adoption and foster policies, and a budget. Most are staffed by volunteers. It is usually after individuals within the organization have to provide medical care for the animals that the veterinarian with a special interest in that species is approached. The individual veterinarian needs to decide if her/his role will be one of volunteer pro bono, per-diem surgeon, as-needed or regular consultant, and/or board member. In any role, it usually is a labor of love; working with a shelter takes time and commitment.

The first step is usually to negotiate reduced fees for animals owned by the shelter, although most soon realize that a knowledgeable veterinarian can assist not only with the medical care of the animals, but help the board with many aspects including liability and insurance needs (largely due to injuries volunteers or visitors may sustain while handling the animals), dietary and husbandry requirements, preventive health care including parasite control, vaccinations, dental care, among other things. If appointed or elected to the board of a shelter, the veterinarian can also serve as a medical director and set up programs to prevent disease outbreaks as well as training of volunteers to recognize illness. The veterinarian may be crucial to assisting a shelter establish record keeping, particularly of intake, adoption, foster, and medical needs. Vaccination records also must be kept. Many shelters rely on volunteer's memories rather than establish an electronic database and records system.[12] Numbers of animals on-hand at the start and end of the year, numbers of adoptions, fostered, transferred, died, etc. are statistics that are needed not only to create a budget plan for the shelter, but are needed to apply for many types of grants.

In determining compensation, most shelters have extremely limited funding. Providing at-cost medications, supplies, diagnostic testing or at a reduced rate, with minimal fees for actual labor are usually what is negotiated. Actual fees, costs, and discounts need to be established in writing, with a clause to review the fees at least on an annual basis. Establish which shelters, sanctuaries you will work with, and set a limit. It is not unusual for word to get out that you do a service for one shelter, and others try to enlist your services as well. In urban areas, a practitioner could spend all clinical time just working with shelter animals, but economically this is not realistic. Set special appointment or visitation times for shelter animal examinations, procedures and surgeries so that you are not taking time away from regular clinic duties. The veterinarian's duties and services to be provided to the shelter in any capacity should be formalized in writing. This is extremely helpful as volunteer organizations have frequent turnover of people involved.

The veterinarian also may assist in the development of contracts with municipal animal control facilities and humane animal facilities that do not accept the avian/exotic species. This may include facilities inspections from the municipal animal control agency before they will transfer an animal to the private shelter. The veterinarian also is responsible for making sure that animals adopted out from the shelter conform to local animal regulations. Spay or neuter, vaccinations and microchipping are frequent requirements, with documentation provided to the new owner. The veterinarian also is responsible for issuing rabies vaccination certificates and other health certificates for transport as required locally.

When working with any animal rescue organization, there will be a large turnover of volunteers and even board members. Many caregivers will experience compassion burnout, particularly if they have fostered many terminally ill and geriatric animals. The veterinarian can provide assistance and training to shelter volunteers to deal with animal death. Grief counseling can be made available.

General guidelines

The veterinarian needs to make sure the animal care is appropriate for the species being sheltered including psychological needs. Animals arriving at shelters are stressed, many are in sub-optimal nutritional and health status, and may experience separation anxiety from their former owners. Volunteers accepting surrendered animals can follow an intake checklist (Table 2) that ensures an animal in medical need is seen by the veterinarian. This checklist (Table 2) is an intake form in use by the Washington Ferret Rescue & Shelter (WFRS) in Kirkland, Washington State and can be adapted for other shelters and species. In areas with endemic heartworm, heartworm diagnostic testing and preventive medications should be included in an intake evaluation. Some ferret facilities also require Aleutian Disease titers and/or Canine Distemper titers as part of admission process. The veterinarian is instrumental in establishing the needed evaluations for a specific shelter. The veterinarian also trains volunteers in accomplishing the checklist. Isolation and quarantine procedures and facilities should be in place for animals newly surrendered. Each animal admitted to the shelter should have a file that includes the surrender papers, history information, intake record, and medical records. In addition to the stored paper records, computerized record-keeping should be encouraged. When an animal is adopted, fostered, or dies, this information is also kept in that animal's record. Storage requirements for these records should be determined by the Board and may comply with medical records retention requirements in that locality.

The veterinarian assesses the animals and can make the recommendation if the animal is adoptable or should be fostered by a volunteer of the shelter. The shelter needs to set the parameters and training programs that qualify a volunteer to be a fosterer. Volunteering to be a "fosterer" should not be a way to get a free exotic animal. It is recommended that a contract between the shelter and the trained fosterer be signed to ensure the fostered animal returns for regular medical care and veterinary attention. Some shelters that do not have permanent facilities put all animals into foster care defined as housing until a permanent home and owners can be found. Other shelters that have facilities to house animals deemed adoptable (i.e., good temperament, at most middle-aged, and healthy). Animals defined that need to be fostered include behavioral problems, geriatric condition, or continuing medical needs. With this definition, foster care is essentially "pawspice care" and necessitates the shelter maintain frequent contact with the foster caregivers to ensure the health care and well-being of those animals. It is recommended that a contract between the shelter and the trained fosterer be signed to ensure the fostered animal returns for regular medical care and veterinary attention or is still available for adoption where there is no shelter facility. Fostered animals or any being held for adoption all legally belong to the shelter and therefore it is necessary for the shelter to track them, maintain records, and ensure health and well-being. Samples of adoption and fostering agreements are presented in Tables 3 and 4.

The veterinarian can also make recommendations on housing, disinfectants and cleaning methods used, protocols for sanitation, appropriate disposal of waste, appropriate storage of food and other products needed for the care of the animals. The facility and procedures should emphasize stress and noise reduction, no crowding, and no exposure to potential predators.

Table 2. Sample shelter intake record.

WASHINGTON FERRET RESCUE & SHELTER
SHELTER INTAKE RECORD

Ferret Name: ____________________________ **Sex** (M/F): ________ **Date of Arrival**: _______________
Age: ____________________ **Ear Tattooed**? Yes No *(description)*: _______________________________
Color: *sable dk /lt sable hvy/lt silver champagne chocolate albino DEW black (solid) cinnamon*
Markings: *mitts bib kneepads panda hood blaze roan*
Nose: *pink speckled spots black outlined* **T** *on nose*

Intact (Y / N): ________ **Scent Glands** (Y / N): ____**Microchip** (Y / N): ____________________________
Weight on Arrival in grams: ________________**Underweight**: (Y / N) @ 1 week: @ 2 weeks:
Type of food ferret is eating if known:
____

Intake Care (check upon completion):
☐ Nails Clipped
☐ Ears Cleaned *(note anomolies below)*
☐ Ivermectin Injection *(0.2cc)*
☐ Bathed *(if necessary)*
☐ Advantage Flea Control - *2 drops between shoulder blades or Revolution*
☐ Teeth (note condition) ____
☐ Distemper (date) ______ **0.5cc Children's Benedryl 20 mins prior* 2nd Distemper (date):*__________
☐ Imrab-3 Rabies (date) **0.5cc Children's Benedryl 20 mins prior**
☐ Fecal Float Result ________________ Date _____________

Physical Inspection:
☐ No Apparent Physical Problems
☐ Inflamed Prepuce ___________________
☐ Swollen Vulva ____________________
☐ Hair Loss ____
☐ Scars/Marks/Scabs
☐ Other ____

Temperament:
☐ Good temperament
☐ Aggressive
☐ Seems to dislike other ferrets
☐ Lap Ferret
☐ Biter

Condition on Arrival:
☐ Excellent
☐ Good
☐ Fair
☐ Poor

Surrendered by:
☐ Owner
☐ Animal Control
☐ Humane Society
☐ Other ____

Shelter volunteer(s) doing intake:
____
Request for Vet Visit (Y/N): __________________ Date of Vet Visit: ___________________________
Adoptable? Yes No **Foster?** Yes No Temp **Entered in Computer?** Yes No

Table 3. Example contract for adoption.

WASHINGTON FERRET RESCUE & SHELTER
CONTRACT FOR ADOPTION

I. The ferret(s) will be handled in a loving, gentle manner by all members of my household and by any guests that visit my home. The ferret(s) will be fed quality dry ferret food, have access to fresh water, will be kept indoors and will be housed in a cage or ferret-proof room when no one is at home to supervise the activities of the ferret(s). The ferret will be seen regularly by a veterinarian knowledgeable about ferrets.

II. If adopters are unable to spend quality time with the ferret(s), unable to keep them as pets, neglect, abuse, or mistreat the ferret(s), or must give them up for any other reason, the ferret(s) will be returned to, or reclaimed by, the Washington Ferret Rescue & Shelter (WFRS). A full refund of the adoption fee is guaranteed up to 7 days from the adoption date.

III. If the ferret's health is ever in question, the adopter will seek veterinary care. It would be appreciated if you would contact the WFRS if any health problems arise or when the ferret passes on so we may better our understanding of breeding, health, and life spans of ferrets. Ferrets will be vaccinated yearly against Canine Distemper or as antibody titer drops and it is required by State Law to have the ferret(s) vaccinated against Rabies once a year as required by regulations in the location of the adopter's residence. The next Canine Distemper shot due:______ _____The next Rabies shot is due:_______________________

IV. Adopter understands that the WFRS cannot and does not make any presentations or guarantees, either expressed or implied as to the temperament, habits, or background of the named ferret(s). Therefore, in consideration of receiving the ferret(s), I hereby release the WFRS from any and all claims of liability for the actions of the ferret(s) being adopted.

This agreement applies to the following ferret(s):

____________________ __________________ _________________ __________________

Signature:____Date:____________

Volunteer Signature:______________________________________Date:____________

Questions? Call or email:____

A copy of this contract will be kept on file at the WFRS.

Table 4. Example contract for fostering.

WASHINGTON FERRET RESCUE & SHELTER
CONTRACT FOR FOSTERING

I. The ferret(s) will be handled in a loving, gentle manner by all members of my household and by any guests that visit my home. The ferret(s) will be fed quality dry ferret food, have access to fresh water, will be kept indoors and will be housed in a cage or ferret-proof room when no one is at home to supervise the activities of the ferret(s).

II. If adopters are unable to spend quality time with the ferret(s), unable to keep them as pets, neglect, abuse, or mistreat the ferret(s), or must give them up for any other reason, the ferret(s) will be returned to or reclaimed by the Washington Ferret Rescue & Shelter (WFRS).

III. If the ferret's health is ever in question, the fosterer will contact WFRS. The WFRS will direct the Fosterer to a WFRS-approved veterinarian or emergency facility. The WFRS may or may not authorize funds if the fostered ferret is taken for medical care without WFRS being notified for approval of costs otherwise the Fosterer must pay for veterinary care. The Fosterer understands that the WFRS cannot and does not make any presentation or guarantees, either expressed or implied, as to the temperament, habits, or background of the named ferret(s). The Fosterer must bring the fostered ferret(s) in to the WFRS at least on an annual basis for medical examinations, regular treatments and vaccinations. The WFRS expects the Fosterer to attend training sessions as they are offered at the WFRS. The Fosterer understands that the ferret(s) legally belong to the WFRS which is responsible for their health and well-being.

Therefore, in consideration of fostering the ferret(s), I hereby release the WFRS from any and all claims of liability for the actions of the ferret(s) being fostered.

This agreement applies to the following ferret(s):

________________ ____________________ __________________ ________________

Signature:____Date:_____________

A duplicate copy of this contract will be kept on file at the WFRS

Specific Species Shelters: Special Considerations for Health Management[5-9]

Depending on endemic disease problems in a geographic area, and whether animals are predominantly coming from indoor or outdoor homes, backyard breeders or pet stores, the following lists of potential disease problems within a shelter will vary. The following are guidelines.

Ferret shelters

Anorexia with/without gastrointestinal ulceration and subsequent diarrhea or hemorrhage, just due to the stress of being surrendered along with changes in the food or habitat is often termed "shelter shock". The ferret suffers severe weight loss during the first week of arrival, despite all other health parameters being normal. These ferrets can be placed on famotidine at 2.5 mg/kg PO q24h (Pepcid Complete, Consumer Pharmaceuticals Co., Fort Washington, PA), and weighed daily. Hand-feeding soup (Carnivore Care, Oxbow Pet Products, Murdoch, NE) is used to nurse them through the first few days. In some cases, the ferret will be taken home by one of the volunteers to nurse through this period and help it become adjusted to its new life. Newly arrived ferrets are kept separate from the main colony until the medical issues listed below have been addressed and the ferret seems to be adjusted to shelter life. Attempts at this time are made to find the ferret suitable cage mate(s) or play groups for socialization.

Communicable disease and parasitic problems that are of particular importance in a ferret shelter include fleas, earmites, heartworm, coccidia, *Cryptosporidium*, influenza, SARS-CoV-2, canine distemper, and possibly enteric bacteria such as *Salmonella* or *Campylobacter* sp. Occasionally fungal skin infections may be found. Rabies in some locales may be considered if the ferret has been outdoors.

Non-communicable disease health concerns: anorexia, gastrointestinal ulceration, diarrhea, weight loss: stress, diet, and husbandry changes, geriatric conditions including heart disease, neoplasia, dental disease, trauma, anal/rectal complications following anal sacculectomy. If unneutered, surgical neutering must be done prior to adoption or fostering. Anal sacculectomy can be considered an elective surgery, but one that may increase the adoptability of the ferret. Most ferrets over the age of 4 years will need attention for adrenal disease treatment.[9]

Rabbit shelters

Many rabbits are brought to shelters that are feral (i.e. were once pets but were turned loose frequently in urban parks). These rabbits may be extremely anxious and frightened in addition to facing diet changes, caging of some type, and human handling. Use of anti-anxiety medications such as midazolam (Midazolam, Baxter Healthcare Corp., Deerfield, IL) when these rabbits are first brought to the shelter may help the rabbit adjust. Many of these rabbits are intact, may be pregnant or lactating and reproductive status must be included in the initial physical examination. Rabbits surrendered should be kept separate from the main shelter until disease status is determined and the rabbit is eating well with normal stools. The first few weeks a rabbit is admitted to the shelter it should be closely observed for urination, defecation, weight changes, and signs of stress. Upon arrival, the rabbit should be weighed and an individual intake record should be started. Full evaluation of each rabbit may be delayed for a few hours after transportation to allow the rabbit to become less frightened. Feral rabbits captured from outdoors may need to be kept in cooler rooms than rabbits accustomed to indoor home temperatures. Rabbit shelter volunteers should strive to prevent abrupt and loud noises as these may severely frighten the already tense rabbits.

Husbandry concerns include caging areas with solid footing rather than flooring entirely of bars or mesh. Substrates such as recycled newspaper pellets may make sanitation easier, although all wet and soiled bedding must be removed from the caging on a daily basis. Various commercial rabbit pellets, fresh hay, herbs, and greens must be stored to prevent spoilage and access by insects or rodents. Ventilation of any housing area should allow fresh air and circulation adequate to prevent ammonia and dusts from accumulating. Rabbits noted with upper respiratory signs should be housed separately from the resident shelter animals. Respiratory disease problems will be exacerbated by stress and introduced rabbits may potentially carry bacterial strains different from the resident rabbits. Shelters may require *Pasteurella* serology and/or culture and treat individual rabbits and determine appropriate adoption status. Surgical neutering should be delayed until respiratory signs are cleared, and the rabbit has adjusted to its new housing and diet. Adoption criteria include temperament, age, and existing medical conditions.

Communicable disease and parasitic problems that are of particular importance in a rabbit shelter: fleas, cuterebra, ear mites, fur mites, coccidia, *Pasteurella* sp., *Treponema*, *E. cuniculi*, dermatophytosis, abscesses.

Non-communicable disease health concerns include stress, anorexia, enteropathy, dental disease like malocclusions and abscesses, urinary calculi, sore hocks, dermatitis secondary to urine scalding, fur matting, other illnesses or injury such as trauma, wounds, and fractures. A rabbit presented with head tilt will need to be evaluated for otitis interna or neurologic problems due to multiple etiologies.

Other exotic mammals

Small mammal shelters are usually individuals or clubs organized by fanciers of specific species. Among the more common surrendered small mammals are guinea pigs, rats and sugar gliders, although a smaller number of hamsters, gerbils, fancy mice, and hedgehogs may also be abandoned at municipal animal control facilities, pet stores, or just turned loose to fend for themselves. All of these animals have unique diet, husbandry and medical needs. In the author's experience, most of these rescued animals are cared for in individual homes rather than designated shelter facilities. With all the rodent species, rescuers need to be aware of the potential to develop severe allergies to rodent antigens and should be encouraged to minimize dusts, exposure to urine/feces, and consider ventilation and filtration systems for rooms housing more than two or three rodents. Rodents should be housed in species-specific rooms and not mixed. The veterinarian may consider multiple-animal fee rates and care plans based on previous working relationships with these individuals. Table 5 lists medical conditions that need to be considered with these rescued species. All stressed rodents may develop enteropathies and have fatalities due to *Clostridium piliformes* (Tyzzer's disease) or *Clostridium perfringens*. Rodents may also carry the zoonotic tapeworm, *Rodentolepsis* (*Hymenolepis*) *nana*. With rats, and to a lesser extent mice, *Streptobacillus moniliformes* is usually carried, and can cause rat bite fever in humans, which can be extremely serious.

Table 5. Species-specific rescued animal concerns.

Animal	Concern
Guinea pigs	Anorexia (stress, change in diet, environment), enteropathy, hypovitaminosis C, dental disorders, dermatophytosis, pododermatitis including overgrown toenails, lice, respiratory disease (*Streptococcus* sp., *Bordetella bronchiseptica*, *Chlamydophila caviae*, etc.); urolithiasis (cystitis). Guinea pigs do not tolerate changes readily, and many rescued cavies require hand-feeding and nursing care to adapt them to a new home.
Rats	*Mycoplasma pulmonis* (other pathogens within the chronic murine respiratory complex, including Sendai virus, Sialodacryoadenitis virus; CAR bacilli, *Corynebacterium kutscheri, Streptococcus pneumoniae, Bordetella bronchiseptica*); dermatophytosis, mange mites, enteropathies (frequently dietary changes), pinworms, neoplasia.
Hamsters	Trauma including wounds, fractures, demodectic mange mites (immunosuppression, stress), neoplasia, enteropathy especially with *Clostridium perfringens*. Cheek pouches should be checked for impaction, infection, neoplasia. Hamsters that have been neglected may be malnourished, torporous, hypothermic, and dehydrated. Hamsters should be housed individually. Dwarf hamsters have different dietary requirements than Syrian (golden) hamsters, although other types of environmental requirements are similar.
Gerbils	Malnutrition, pneumonia and skin conditions such as dermatophytosis and nasal dermatitis related to high humidity (urine, feces, dripping water bottles and other sanitation issues), degloving injuries to tail from improper handling, enteropathy, trauma including fractures.
Hedgehogs	Mange mites, enteropathy, pododermatitis (poor sanitation and accumulation of feces), torporous due to cool ambient temperature, dehydration, malnutrition, neoplasia, pneumonia, ataxia associated with condition currently called wobbly hedgehog syndrome. Many hedgehogs initially may have been obese and when they are abandoned or face a major dietary change may become anorexic with secondary hepatic lipidosis.
Sugar gliders	Malnutrition, osteomalacia/fibrous osteodystrophy, torporous due to cool ambient temperature, dehydration, trauma (wounds, fractures), enteropathy particularly due to *Giardia* sp., pouch infections

Herpetologic shelters

Reptile rescues may range from individuals housing all the animals within their homes to those who can build a free-standing reptile sanctuary complete with a rainforest room. Most also utilize individual foster families as housing for adoptable or hospice reptiles. Foster families apply to the program and undergo training on recognizing medical problems as well as proper husbandry for the species selected to foster. Often they are provided with caging and lights appropriate to the selected species from the rescue organization. Once an animal is fostered there are quarterly examinations made by the adoption committee until the animal is adopted or deceased. Half-way homes also serve as quarantine facilities since newly acquired reptiles are segregated from other rescued reptiles. Unfortunately, a foster home system restricts the number of animals that can be accommodated. Shelters frequently stop accepting mature iguanas due to the large number of iguanas needing to be placed and the small number of adoptions of mature iguanas.

Reptiles surrendered to shelter and rescue organizations frequently fall into two categories: those that are too large/too aggressive to keep and those that have medical conditions created by poor husbandry. Often owners of reptiles are ill-equipped, particularly financially, to properly care for these animals. Medical and behavioral evaluations are important in determining immediately adoptable, adoptable after medical treatment/socialization, permanent fosters/hospice animals and those that should be humanely euthanatized. Legal issues surround venomous snakes and lizards, crocodilians and snakes over 12 feet in length in many jurisdictions. In many states/municipalities these animals are subject to confiscation by state/municipal animal control personnel. While some jurisdictions will require euthanatization of these animals, others place them in sanctuaries with people knowledgeable on how to properly care for these species.

All newly acquired reptiles should be given a physical examination by trained personnel. Any medical conditions are noted and the foster family is authorized to seek veterinary care and treatment from local veterinarians experienced with reptiles, that the organization/rescue has pre-approved. Most medical conditions seen by the author involve inadequate uVB light exposure, imbalanced nutrition, hypothermia and internal parasitism. Once improper husbandry and diet practices are corrected and parasitism addressed, these reptiles are usually adoptable. An exception may be reptiles infected with *Nannizziopsis* sp., boid inclusion disease, or *Cryptosporidium*, as these may pose a risk to other reptiles within a home or facility.

Those animals that are deemed too aggressive can be fostered with specially trained volunteers for socialization. Sometimes, retraining is effective but often these animals are deemed too dangerous and euthanized per shelter/sanctuary policy. In an effort to ameliorate hormonal aggression and subsequent abandonment, the author knows of one rescue organization that sponsors a spay/neuter program for iguanas. This program has decreased the number of iguanas seeking placement in that facility.

Reptile specific veterinary concerns include internal/external parasites (particularly in wild-caught animals), anorexia (stress, new food items, changes in environment), hypocalcemia/hyperphosphatemia (renal disease, metabolic bone disease), hepatic lipidosis (carnivorous lizards), bacterial abscesses, sepsis, respiratory disease in chelonians (e.g., *Herpesvirus*, *Mycoplasma* sp., *Bordetella bronchiseptica*), dystocia, neoplasia. Chelonians should be isolated from snakes because of possible *Entamoeba* cross-infection.

Avian Rescues and Sanctuaries

There are many well-known avian sanctuaries and rescue organizations across the United States. Veterinary concerns for avian rescues include quarantine procedures for new acquisitions, general medical screenings, geriatric screenings, disease outbreak protocols, nutritional requirements of the species housed as well as individual medical attention. Even minimally funded shelters need to have quarantine procedures, excellent sanitation protocols, predator resistant caging and adequate nutrition.

Frequently parrots are given up because of behavioral issues (e.g., screaming, aggression, feather picking). Behavioral assessment is very important, especially in determining adoptable versus sanctuary (unadoptable) candidate. Most birds surrendered come from single bird households, and many have underlying medical problems as well.

Specific veterinary concerns include internal parasites, viral infections/exposure (circovirus, bornavirus- PDD, polyoma, herpesviruses), respiratory disease (*Aspergillus*, *Chlamydophila*), hypovitaminosis A and dietary insufficiencies, dystocia, hepatic lipidosis, cardiac disease, neoplasia, and psychological disorders.

A recent tragedy occurred at a local avian rescue. It was a large facility housing 400 cockatoos, macaws, African greys, and Amazons. It was located at the founder's house, and the website as well as on tours and with contacts, and house-call visits by avian veterinarians over the years, it was promoted as being a 501(c)(3) facility. The public was not allowed on the premises; surrenders would be left at the gate. They used pheasants in the larger flight cages to pick up food detritus and act as rodent control. All birds were housed outdoors even though some needed winter shelter. The founder apparently was turning away offers of help with animal care recently, and when contact was lost, she was found deceased in her home, and all but 1 of 400 birds were dead as well.

Acknowledgments: The following organizations are acknowledged: Washington Ferret Rescue & Shelter, Kirkland, WA; Best Friend Rodent Rescue, Everett, WA; Oregon Tiger Sanctuary, Eagle Point, OR; Pacific Primate Sanctuary, Haiku, Maui, HI.

References

1. Kreisler RE, Spindel ME, Rishniw M. Surveys of salary, benefits, and job responsibilities for veterinarian employed in the field of shelter medicine in the United States conducted in 2011 and 2018. Top Companion Anim Med. 2020;39:100430:1-11.

2. Powell L, Reinhard CL, Serpell J, Watson B. A survey of veterinary student and veterinarian perceptions of shelter medicine employment. J Vet Med Educ. 2021;e20210112. doi:110.3138/jvme-2021-0112.

3. Powell L, Reinhard CL, Serpell J, Watson B. Workplace relations and opportunities for career development impact the retention of veterinarians in shelter medicine. Front Vet Sci. 2021;8:732105. doi:10.3389/fvets.2021.732105. eCollection 2021.

4. Laderman-Jones BE, Hurley KF, Kass PH. Survey of animal shelter managers regarding shelter veterinary medical services. Vet J. 2016;210:68-76.

5. Association of Shelter Veterinarians. Available at: https://www.sheltervet.org. Accessed March 31, 2022.

6. Hanley, CS. The care of small mammals in the animal shelter. In: Miller L, Zawistowski S, eds. Shelter Medicine for Veterinarians and Staff. 2nd Ed. Ames, IA: Wiley-Blackwell; 2013:185-200.

7. Nicholson DA. Avian care in the animal shelter. In: Miller L, Zawistowski S, eds. Shelter Medicine for Veterinarians and Staff. 2nd Ed. Ames, IA: Wiley-Blackwell; 2013:225-246.

8. Isaza N, Isaza R. Reptile and amphibian care in the animal shelter. In: Miller L, Zawistowski S, eds. Shelter Medicine for Veterinarians and Staff. 2nd Ed. Ames, IA: Wiley-Blackwell; 2013:201-224.

9. Johnson-Delaney CA. Rescue centre medicine. In: Johnson-Delaney CA, ed. Ferret Medicine and Surgery. Boca Raton, FL: CRC Press; 2017:437-448.

10. University of California Davis Koret Shelter Medicine Program. Shelter Medicine 101 VME 424-01. Available at: http://www.sheltermedicine.com. Accessed April 3, 2022.

11. Kirk JT. Spock quote. Wrath of Khan. Star Trek II [Film]. Hollywood, CA: Paramount Pictures; 1982.

12. Vinic T, Dowling-Guyer S, Lindenmayer J, et al. Survey of Massachusetts animal shelter record-keeping practices in 2015. J Appl Anim Welf Sci. 2020;23(4):385-401.

Section 37

Technical Topics

Christina Miller, RVT, BSc;
Elizabeth Vetrano, RVT
Session Chairs

Food, Function and Fecals: Gastrointestinal Anatomy, Physiology and Nutrition of Rabbits, Guinea Pigs and Chinchillas

Micah Kohles, DVM, MPA

Session #2655

Affiliation: From University of Nebraska-Lincoln, Lincoln, NE, Oxbow Animal Health, Murdock, NE 68407, USA, Woodland Animal Hospital, Hickman, NE 68372, USA.

Introduction

The gastrointestinal (GI) function of rabbits, guinea pigs, chinchillas, and other exotic companion mammal (ECM) species is complex and dependent on proper nutrition. Understanding the GI anatomy and physiology of ECM species is key to properly educating owners, in addition to treating and diagnosing many common diseases as most we see in clinical practice are directly or indirectly related to nutrition. In this lecture we focus on the most common ECM species and discuss their GI tracts and what diets are on the market including the good, bad and ugly.

Rabbits

Domestic rabbits (*Oryctolagus cuniculus*) are strict herbivores, classified as hindgut (cecum and colon) fermenters, concentrate selectors, and are adapted to ingest large amounts of high fiber food. As prey species, maintaining a small body size is advantageous. Their unique GI system, with 12- to 16-hour GI transit time and ability to differentiate particulates of fiber, allows them to remain small and active, while surviving on high-fiber, low-energy density plant materials. GI disease in this species is a very common presenting issue in clinical practice. It is often associated with dysbiosis and can present as both primary and secondary disease.

They are mostly crepuscular and nocturnal feeders. They are anatomically and physiologically adapted to handle significant amounts of low-energy density fibrous food and to effectively utilize the nutrients found in a high-fiber diet; however, due to their small body size, they are unable to store large amounts of food material and therefore utilize a unique process (wash back colonic separation mechanism) to separate fiber based on size. The easily fermentable component of the diet is retained, while the slowly fermentable components of the diet (preeminently cellulose based plant fiber) are rapidly eliminated. This separation process allows for the production of two types of stool: a dry, hard, high-fiber fecal pellet and an enlarged, soft fecal pellet that is covered in mucous and referred to as a cecotroph (cecal, night stool, etc.) which is immediately ingested. It should be noted that when rabbits ingest cecotrophs, normal mastication does not occur and the cecotrophs are swallowed intact. Rabbits, like all herbivores, lack mammalian enzymes to break down the cellulose components of their plant-based diets and thus have a symbiotic relationship with a diversity of gut microbes and flora. Some of the most common issues with domestic rabbits, both in companion and commercial animals, are related to the GI system. The symbiotic relationship rabbits have with the microflora in the cecum allows them to effectively ferment structural plant carbohydrates which serves as their main energy source.

Rabbits have a large cornea and laterally placed eyes, which give them a panoramic field of view when grazing and foraging. Food is selected based on smell and tactile perception of the lips, as their visual field does not include the area just under their nose. As concentrate selectors, rabbits will choose the most tender, succulent plant parts which are nutrient dense and lowest in fiber to meet their nutritional needs, owing to a high metabolic rate. Rabbits have hypsodont or open-rooted (constantly growing) incisors and cheek teeth (premolars and molars).

Rabbits' incisors are designed to apprehend and cut through vegetation. The large upper incisors have two smaller incisors behind them (peg teeth). The occlusion of the upper and lower incisors allows the teeth to be constantly sharpened. The premolars and molar teeth, also known as cheek teeth, are designed to process hard, fibrous foods. The fibrous food is ground between the cheek teeth with jaw movements up to 120 per minute. This repetitive, rapid chewing movement helps to maintain proper dental health. If low-fiber mixed cereal or seed-based foods are fed with minimal or no hay, the rapid chewing activity is no longer required, which can result in dental disease, such as malocclusion and root elongation, as well as GI issues. As the teeth elongate secondary to inadequate wear, the jaws are forced apart and the radiographic appearance of the diastema increases. Over time, the jaws begin to no longer occlude properly, and mandibular movement is reduced. The condition deteriorates as the teeth become mobile and maloccluded, resulting in the development of spurs that may lacerate and ulcerate the tongue and buccal cavity. A suitable diet of high-fiber hay and high-fiber, uniform pellets will help control the growth of the teeth and prevent overgrowth and malocclusion. As concentrate selectors, rabbits, like deer, have a predisposition to "select" the lower fiber, more tender portions of their diets which, if they are fed cereal based or mixed pellets, can lead to selective feeding and malnutrition.

The stomach is J-shaped and large compared to most monogastrics. It comprises roughly 15% of the GI tract and has both a well-developed cardiac sphincter that prevents vomiting and a muscular pyloric area. The pH of the stomach is low and ranges from one to six. The pH is affected by a variety of factors, including the site of determination, the presence or absence of cecotrophs, the duration of time from feed intake to measurement, and the age of the rabbit. The pH of weanling rabbits is higher (5-6.5), which permits the passage of bacteria through the stomach to the hindgut to colonize the cecum.

The small intestine is made up of the duodenum, jejunum (longest), and ileum, is short compared to other species at ~3 meters in length and comprises ~12% of the total GI tract. Transit time through the jejunum and ileum is 10-20 and 30-60 minutes, respectively. The duodenum is the largest of three segments but shortest in length (~3-5 inches in rabbits). The distal end of the ileum is thickened and referred to as the sacculus rotundus, which is unique to the rabbit and marks the intersection of the ileum, colon, and cecum. This location signals the beginning of the large intestine and links to the ampulla coli, which forms an intersection of the distal ileum, cecum, and proximal colon. Located at the junction of the ileum and the large intestine is the ileocecal valve, which regulates the movement of material from the small intestine into the large intestine and prevents the retrograde movement of intestinal contents.

The small intestine is the site where the greater part of digestion and absorption takes place by passive and active transport throughout the mucosa. Like monogastric animals, rabbit's stomach aids in initiation of digestion through secretion of hydrochloric acid and pepsin. Pancreatic enzyme production, such as amylase, is minimal, with most coming from saliva or bacterial production within the cecotroph. A key function of the small intestine is the absorption of nutrients contained in cecotrophs. Cecotrophs contain essential nutrients, including volatile fatty acids, hydration, protein, vitamins and amino acids which help to meet nutrient requirements. They are coated in a protective mucus which is applied by the goblet cells in the descending colon that protects them from the acidity of the stomach prior to absorption in the small intestine. The transit time of the small intestine is rapid compared to other herbivorous species and fiber is quickly moved to the cecum and large intestine.

The rabbit cecum is very large and has a capacity roughly ten times that of the stomach, making up around 40% of the total GI tract. It is a thin-walled organ that folds on itself multiple times and has an internal surface made up of a long spiral fold (sometime referred to as spiral valve) that is continued into the beginning of the colon, an area known as the *ampulla coli*. The distal tip of the cecum is known as the vermiform appendix. This tube-like area is narrow and thickened compared to the cecum and contains significant amounts of lymphoid aggregates which secrete bicarbonate ions into the cecum. The bicarbonate ions act as a buffering agent for the volatile fatty acids formed during cecal fermentation.

The colon is divided into three sections, the ascending, transverse, and descending colon. Because the *fusus coli* (located anatomically at what is referred to as the transverse colon) forms such a natural division between the morphologically and functionally distinct ascending and descending sections of the rabbit colon, many physiology texts have abandoned the three-section description and simply use the terms proximal and distal colon. The colon is a very functional component of the hindgut and is characterized by sacculations (haustra) and bands (taeniae). The ascending colon contains small protrusions, approximately 0.5 mm in diameter, that are commonly referred to as warzens. These wart-like protrusions are believed to be unique to rabbits. They potentially represent an increase in the surface area of the colon that would favor absorption and also may assist in mechanical separation of gut contents. The *fusus coli* has a mucosa that is four to five times thicker than the descending colon and contains ganglion aggregates. The *fusus coli* is commonly referred to as the pacemaker of the hind gut and controls the retrograde and normograde peristaltic activities that occur during the formation of soft and hard feces. The descending colon's mucosa is smooth, and while it contains numerous goblet cells, does not contain obvious specialization and usually contains hard fecal pellets.

There is an almost continuous flux of material between the cecum and the proximal colon. The mixture of fluid and nutrients is moved through vigorous peristalsis in both normo- and retrograde directions. These contractions result in large indigestible fibrous particles accumulating in the center of the colon where they are rapidly transported along the colon to the rectum for defecation, usually within four hours of ingestion. Conversely, smaller, more fermentable fibrous particles accumulate at the periphery of the colon where through retrograde peristaltic contractions of the haustra, they are returned to the cecum for further fermentation.

Periodically there is a dramatic change in the peristaltic activity of the cecum and colon as the retrograde movement of smaller fiber particles ceases and a large quantity of the cecal content is expelled into the colon. These contents are excreted, usually once to twice daily, as soft, mucous covered cecotrophs. This process is controlled by the *fusus coli*. Because rabbits do not completely ferment fiber, the ability of the colon to rapidly eliminate large indigestible fiber particles and retain smaller, more digestible fiber for further fermentation, makes the rabbit an extremely efficient herbivore, capable of surviving on very low-quality forage.

The cecotroph fermentation process depends heavily on an appropriate diet and the action of resident bacteria (especially *Bacteroides* spp. and *Clostridium* spp.) and protozoa which are vital to the GI health of the rabbit. The diversity of bacterial flora and other organisms within the cecum is substantial and includes many bacterial species dependent on location within the cecum lumen or wall, unidentified anaerobic species, ciliated and flagellate protozoa, and a rabbit specific yeast. The effects of bacterial fermentation include the production of volatile fatty acids (VFA) including acetic, formic, propionic, and butyric acids. These are a major energy source for rabbits. A portion of these VFAs are contained in cecotrophs and utilized upon re-ingestion, while another portion is absorbed directly across the cecal mucosa.

There is still much to learn and understand about the microbial populations of the rabbit hindgut. Fermentation produces volatile fatty acids, vitamins B, C, and K, and amino acids which are eventually digested in the small intestine after cecotrophy, which occurs once or twice a day, usually at night. Over one hundred additional strains of anaerobic bacteria have been isolated from the cecum mucosa of rabbits, and of those, very few have been identified.

In healthy and physically able rabbits, the large, soft mucous-covered cecotrophs are ingested directly from the rectum (midnight snack). Multiple factors, including stimulation of rectal mechanoreceptors, perception of the cecotroph odor, and the blood concentrations of various metabolites and hormones, all stimulate the rabbit to directly ingest the cecotrophs. Higher-fiber diets increase cecophagy, whereas high-protein and low-fiber diets decrease cecophagy. Once eaten, the acidity of the stomach breaks down the protective mucous covering of the cecotroph. Continued digestion occurs in the small intestine where cecotrophs represent an important component of the diet providing protein, water and vitamins.

Guinea Pigs

Guinea pigs (*Cavia porcellus*) are obligate herbivores and have a large tongue and narrow oral cavity. Dental disease is a common malady, similar to rabbits and chinchillas, and proper diet (specifically, ample fiber intake) is key to decreasing the likelihood of dental and GI issues. Guinea pigs are hindgut fermenters and are coprophagic. While the belief is that coprophagy is an important function in guinea pigs, its contribution to overall nutritional needs has not been fully characterized. Coprophagy in rabbits is a key source of B vitamins and a means of optimizing the availability of proteins; however, there is uncertainty if guinea pigs are truly cecotrophic, as unlike rabbits, they are not reliant on the key amount of B vitamins found in cecotrophs. Small herbivores utilize one of two primary mechanisms for the colonic separation mechanism necessary for coprophagy. This capability permits rapid passage of less digestible food particles while retaining micro-organisms and more digestible food particles in the cecum, to allow adequate opportunity for fermentation. Rabbits utilize a multi-directional peristaltic motion to separate components of fiber based on size and mix these with other ingesta, bacteria, and water to concentrate more fermentable particles and eliminate larger particles of fiber. Guinea pigs and chinchillas, on the other hand, utilize a process sometimes referred to as "mucus trap," in which bacteria from the cecum are trapped in the mucus of the colon with few to no food particles and returned to the colon through antiperistalsis. The mucous trap process is less efficient than the mechanisms utilized by rabbits. Guinea pigs and chinchillas have a colonic furrow in the ascending colon, in which the concentration of bacteria and nitrogen is twice as high as in the lumen.

The stomach of guinea pigs contains the same four regions, the cardia, fundus, body, and pylorus. The small intestine of guinea pigs is made up of three sections (duodenum, jejunum, and ileum), as found in most species. They are visually non-distinguishable, with the jejunum being the longest section (~90-100 cm). The cecum is the largest component of the guinea pigs' GI tract and contains upwards of 65% of total contents. The flora is similar to the rabbit, being primarily gram positive with anaerobic lactobacilli being the primary organisms in the large intestine. As in the rabbit, a large number of additional organisms including yeast, protozoa, and other bacterial species are prevalent. Gastric emptying time is ~2 hours and total GI transit time is ~20 hours (8-30 hours), and considerably longer (up to 60 hours) if coprophagy is considered.

Chinchillas

Chinchillas (*Chinchilla lanigera*) are obligate herbivores, hindgut fermenters, and cannot vomit, similar to rabbits and guinea pigs. They are similar to the guinea pig than the rabbit in terms of GI anatomy and physiology. Their dentition is also similar to guinea pigs than rabbits, as they have a larger grinding surface owing to a naturally higher fiber and coarser diet than rabbits. As compared to the guinea pig, the jejunum and descending colon are longer. The colon is highly sacculated and the cecum large, yet the cecum holds less contents of the overall large intestine in chinchillas compared to rabbits and guinea pigs.

Nutrition

The discussion of the GI system of herbivorous exotic companion mammals above makes it clear how integral proper diet is to the health of all small herbivores. The keys to a proper small herbivore diet are the following:

1. High-quality high-fiber grass hay (~70-75%).
    - A variety of grass hays should be fed free choice, available at all times. Any high-quality grass hay such as timothy, orchard, meadow or oat is appropriate.
    - Diversity in hay type offering is strongly recommended as this mimics the natural diversity of plants these species would forage for in wild environments.
    - These species will also use hay for bedding and nesting material. It is important to differentiate these uses and ensure the animals always have access to clean fresh hay for ingestion.

2. High fiber uniform pellet (~20-25%).
    - Timothy or alfalfa-based, depending on life stage. Always feed only the recommended amount.
    - It is important to feed only uniform products and avoid products that contain added seeds, fruits, or vegetables, or that utilize additional ingredients (typically starch or carbohydrate based) required for certain manufacturing techniques.

3. Fresh or dried greens (~10-15%).
    - Only appropriate species should be fed, and all should be carefully cleaned. (Organic products are preferred.)

4. Treats (~0-5%).
    - Low-sugar, high-fiber treats.

5. Unlimited fresh, clean water.
    - Multiple sources, such as water bottles and crock, should be available at all times.

High-quality, high-fiber grass hay should be available at all times. Fiber is essential to maintain proper gut health, and to stimulate gut motility (insoluble fiber only). Although there are multiple varieties available on the market, there is minimal nutritional difference between them, and therefore, different species, such as timothy, orchard, oat, and mixed are appropriate. Hay is a product of nature, and therefore, there will be variances in color, texture, feel, and taste. Because of this, it is important to transition between hay bags and offer different varieties of hay. Hay not only provides a key component of the nutritional needs of rabbits and other small herbivores, but it also encourages natural behaviors such as foraging and grazing. Stimulating these natural behaviors may diminish boredom-based behaviors, increase activity, and also provide a sense of security to the animal.

There are many diets available to pet owners, the two most common being seed/cereal-based muesli mixes and hay based, high-fiber uniform pellets. Rabbits, guinea pigs and chinchillas are all concentrate selectors, and if offered a mixed pellet, will selectively feed on certain items from the mix, which will lead to improper and unbalanced nutrition. These mixes are often high in carbohydrates and simple sugars, which can lead to obesity and GI issues, as well as low in fiber, which slows GI motility and decreases normal teeth wear. Uniform hay-based fortified pellets, high in fiber and low in protein, are recommended, as they eliminate selective feeding, thereby guaranteeing a proper vitamin and mineral intake.

Fresh and dried greens are an important and enjoyable component of a rabbit's diet and should be considered, not an option, but an essential part of a healthy animal's diet. They also provide an excellent source of water and enrichment. Varying types and textures is a great enrichment tool. It is recommended to completely wash and clean all greens, and organic products are preferable.

Although treats are not a required component of these animals' nutrition, limited appropriate treats can be beneficial for strengthening the pet/owner bond, training, and enrichment. It is important to limit the number of treats and ensure they are low in protein, fat, calcium, and sugar. Some examples of inappropriate treats are nuts, seeds, popcorn, bread, crackers, and yogurt drops. High-fiber grass and alfalfa-based treats are appropriate in limited amounts, as are small amounts of dried fruits and vegetables. There are far more inappropriate treats on the market, but good products do exist.

Unlimited fresh, clean water is vital, and the water requirements of rabbits have been estimated at 50-100 ml/kg/day depending on their diet. Fresh, clean water should be always available to small herbivores. It is a good idea to offer herbivores water in more than one manner, such as a sipper bottle and bowl/crock, to stimulate increased water intake. Water should be changed daily, as bottles/bowls/crocks will become dirty with normal drinking. Actual water consumption can be significantly affected by diet and other environmental factors.

Recommended Reading

1. Alworth L, Harvey S. Chinchilla anatomy, physiology and behavior. In: Suckow M, Stevens K, Wilson R, eds. *The Laboratory Rabbit, Guinea Pig, Hamster and Other Rodents.* Waltham, MA: Academic Press; 2012:955-966.

2. Bjornhad G, Snipes R. Colonic separation mechanism in lagomorph and rodent species-a comparison. *Zoosystem Evol.* 1999;75:275-281.

3. Brewer N, Cruise L. Physiology. In: Manning PJ, Ringler DH, Newcomer CE, eds. *Biology of the Laboratory Rabbit.* 2nd Ed. Waltham, MA: Academic Press; 1994:63-70.

4. Carabano R, Piquer J, Menoyo D, Badiola I. The digestive system of the rabbit. In: de Blas C, Wiseman J, eds. *Nutrition of the Rabbit.* 2nd Ed. Cambridge, MA, CABI Publications; 2010.

5. Cheeke P. *Rabbit Feeding and Nutrition.* Orlando, FL: Academic Press; 1987.

6. Davies R, Davies J. Rabbit gastrointestinal physiology. *Vet Clin North Am Exot Anim Pract.* 2003;6: 139-153.

7. De Blas C, Wiseman J. *Nutrition of the Rabbit.* 3rd Ed. Boston, MA: CABI; 2020.

8. Franz R, Kreuzer M, Hummel J, et al. Intake, selection, digesta retention, digestion and gut fill of two coprophageous species, rabbits (*Orytolagus cuniculus*) and guinea pigs (*Cavia porcellus*), on a hay only diet. *J Anim Physiol Anim Nutr.* 2010;95(5):564-570.

9. Griffiths M, Davies D. The role of the soft pellets in the production of lactic acid in the rabbit stomach. *J Nutr.* 1963;80:171-80.

10. Hargaden M, Singer L. Guinea pig anatomy, physiology and behavior. In: Suckow M, Stevens K, Wilson R, eds. *The Laboratory Rabbit, Guinea Pig, Hamster and Other Rodents.* Waltham, MA: Academic Press; 2012:575-602.

11. Harkness J, Murray K, Wagner J. 2002. Biology and diseases of guinea pigs. In: Fox J, Anderson L, Loew, Quimby F, eds. *Laboratory Animal Medicine.* 2nd Ed. San Diego, CA: Academic Press; 2002:203-246.

12. Irlbeck N. How to feed the rabbit (*Oryctolagus cuninulus*) gastrointestinal tract. *J Anim Sci.* 2001;79: 343-436.

13. Jenkins J. Feeding recommendations for the house rabbit. *Vet Clin North Am Exot Anim Pract.* 1999;2(1): 143-151.

14. Jilge B. The gastrointestinal transit time in the guinea pig. *Z Verdsuchstierk.* 1980;22:204-210.

15. Quesenberry K, Orcutt C, Mans C, Carpenter J. *Ferrets, Rabbits & Rodents: Clinical Medicine and Surgery.* 4th Ed. St. Louis, MO: Elsevier; 2020.

16. Snipes T, Clauss W, Weber A, Hornicke H. Structural and functional differences in various divisions of the rabbit colon. *Cell Tissue Res.* 1982;225:331-346.

17. Stevens C, Hume I. Contributions of microbes in vertebrate gastrointestinal tract to production and conservation of nutrients. *Physiol Rev.* 1998;78:393-427.

18. Sohn J, Cuoto M. 2012. Rabbit anatomy, physiology and behavior. In: Suckow M, Stevens K, Wilson R, eds. *The Laboratory Rabbit, Guinea Pig, Hamster and Other Rodents.* Waltham, MA: Academic Press; 2012:195-215.

19. Wolf P, Schroder A, Wenger A, Kamphues J. Nutrition of the chinchilla as companion animal-data, influences and independences. *J Anim Physiol.* 2003;87:129-133.

20. Wingerd B. Rabbit anatomy and dissection guide. Eden Prairie, MN: Blue Door Publishing; 2007:57-67.

Food, Function and Fecals: Gastrointestinal Anatomy, Physiology and Nutrition of Ferrets and Rodents

Micah Kohles, DVM, MPA

Session #2901

Affiliation: From University of Nebraska-Lincoln, Lincoln, NE, Oxbow Animal Health, Murdock, NE 68407, USA, Woodland Animal Hospital, Hickman, NE 68372, USA.

Introduction

The gastrointestinal (GI) function of ferrets and other exotic companion mammal (ECM) species is complex and dependent on proper nutrition. Understanding the GI anatomy and physiology of ECM species is key to properly educating owners, in addition to treating/diagnosing many common diseases as most we see in clinical practice are directly or indirectly related to nutrition. In this lecture we focus on ferrets, rodents, hedgehogs and other and discuss their GI tracts and what diets are in the market including the good, bad and ugly.

Ferrets

Domestic ferrets (*Mustela putorius furo*) are members of the order Carnivora, suborder Caniformia, family Mustelidae, subfamily Mustelinae. The scientific name *Mustela putorius furo* roughly translates to mouse eating smelly little thief. There are three species of ferrets (also known as polecats in Europe and Asia): the European polecat (*Mustela putorius*), the Steppe or Siberian polecat (*Mustela eversmanni*), and the black-footed ferret (*Mustela nigrans*). They are related to martens, fishers, otters, wolverines, long tailed weasels, least weasels, mink, badger and skunks. Ferrets are obligate carnivores, meaning they are designed to kill and eat whole prey items, which include all parts of the killed animal. They are specialized, with elongated bodies and short limbs, for the pursuit of warm-blooded prey in holes and tight spaces. Ferrets tend to eat small meals throughout the day and will hide cache excess food in their dens or other hiding places.

Ferrets are obligate carnivores with a very short intestinal tract and are sometimes referred to as super carnivores, because of anatomic, physiologic and behavior traits adapted to a strict carnivore lifestyle. The GI system of a ferret is essentially a long tube from the stomach to the rectum which allows food to pass too quickly for efficient absorption. The entire GI tract is under vagal and sacral innervation. The short length of the overall GI system of ferrets results in a rapid transit time of approximately 3-4 hours.

The gut flora is simple, and therefore GI upset or diarrhea secondary to antibiotic usage is uncommon. Like humans, ferrets are spontaneous secretors of hydrochloric acid which differs from many other predators such as dogs and cats. In addition, they spontaneously produce proteolytic enzymes and the GI tract remains active even under anesthesia. Ferrets do have the ability to vomit, but often do not even in the presence of obstruction or other pathology. Because of their rapid transit time and need for small frequent meals, a prolonged fasting, greater than 3-6 hours, is neither required nor recommended. It should be noted that ferrets with insulinomas may become hypoglycemic much quicker, even less than 3 hours, and therefore require special pre-operative care.

Ferrets have binocular vision, a very broad auditory range, and extremely sensitive olfactory specialization which make them excellent predators. They also have a hinge like temporomandibular joint, large mastication musculature and a large bite force to body mass ratio which allows them to predate larger sized prey.

Ferrets have typical carnivore dentition and oral anatomy with elongation of the carnassial blade, reduction or loss of post carnassial molars, and shortening of the rostrum in comparison to skull length. The upper incisors are slightly longer than the lower incisors when the mouth is closed. The canines are very pronounced, as in other carnivores. The lower premolars as well as first and second upper premolars all have two roots whereas the fourth upper premolars, the carnassial teeth, have three roots. The lower second molar is very small single rooted tooth. Dental formula: I (3/3) C (1/1) PM (3/3) M (1/2) = 34

Ferrets' teeth are precisely interdigitated, which is inline with the bladed shape of their teeth. Their teeth are designed to shear and tear animal tissue and are not designed for grinding. Previous research has pointed to commonly fed extruded diets predisposing ferrets to dental disease or exacerbating what could be consider normal dental wear. It is possible that the shape of their teeth and design for shearing are not in line with the chiseling and grinding process utilized to effectively masticate dry extruded kibble.

The stomach of ferrets is simple and is shaped similar to the dog. The stomach fits in the curve of the liver and has an identifiable pylorus. The two limbs of the pancreas are connected at the midline near the pylorus. It can expand to accommodate large amounts of food even though ferrets typically do not gorge themselves like some other carnivore species. The antrum of the stomach provides the mixing and propulsion of contents for gastric emptying.

The small intestine is quite short (~182-198 cm in length) and consists of three sections all of which contain goblet and villi cells. This short length is one factor that results in such a decreased GI transit time in ferrets. The major duodenal papilla is found ~3 cm from the pylorus and serves as the common opening for the bile and pancreatic ducts. The small intestine is made up of the duodenum, jejunum and ileum. Atropine does not block jejunal motility, as this movement is mediated by hormonal activity. Gastric acid secretion is only reduced by 30% in the presence of atropine. The ileum and jejunum are indistinguishable on gross examination and may be referred to as the jejunoileum which ends at the ascending colon.

The entire large intestine of a ferret, which includes the colon and rectum, is approximately 10 cm in length and lacks both a cecum and ileocolic valve. This junction is inferred by the presence of the anastomoses of the jejunal artery with the ileocolic artery. The ascending, transverse, and descending colon make up the entire colon with the descending portion being the largest.

Nutrition

It is important to look at the evolutionary diet of ferrets to try and better understand their nutritional needs. Although there is no documented proof, it can be assumed that a ferret's diet would reflect that of the polecat. This would be composed of mostly small and medium sized mammals, but could also include a smaller portion of birds, carrion, reptiles, amphibians, eggs, crustaceans, mollusks, insects, worms and fish. Although some evidence reveals ferrets will feed on a small amount of seasonal plant-based foods, their diet is obviously that of animal prey. They are, as we have discussed, strict obligate carnivores.

In view of the previous ferret GI discussion, it is abundantly clear how unique and inefficient their GI tract is and how important feeding an appropriate diet is to their health. To compensate for their inefficient GI system ferrets, require a very concentrated diet which is high in protein and fat as well as low in fiber and carbohydrates. The only carbohydrate sources in a ferret's natural diet are from the gut contents of their prey. Fat, when metabolized, releases twice as much energy as carbohydrate or protein metabolism and therefore should be the main source of calories in a ferret's diet.

While there has been some advancement in ferret diets, the majority of animals are still fed a dry extruded kibble. Most of these diets contain higher than recommended levels of plant based starches which are required for kibble shape and stability as part of the extrusion process. This process consists of finely grinding ingredients and then mixing them into dough which is extruded through a tube under significant heat and pressure. The heating and processing of the animal proteins can potentially make them less digestible. Dry extruded kibbles that contain grains and other plant based starch sources could be related to: 1) Urolithiasis: Due to diets high in plant proteins 2) Insulinomas: Relating to the potential impact of carbohydrates on predilection for the formation this disease 3) Excessive dental wear: Dry extruded kibble has been shown to cause an excessive amount of dental wear in relation to semi moist or natural diets.

Unfortunately, there are limited feeding options currently available to ferret owners. Although feeding a whole prey or raw diet is not owner-friendly, this is the best possible alternative for providing a complete natural diet for ferrets. It is important to educate owners about the potential risks of raw diets, such as bacterial (Salmonella, et.) and parasitic contamination which can be a zoonotic issue. There are multiple semi moist diets on the market as well which, if they fit the nutritional profile, might be applicable but could predispose the ferret to periodontal disease if proper dental care is not maintained.

Unlimited fresh, clean water is vital as ferrets require approximately three times as much water as dry matter. It is a good idea to offer ferrets water in more than one manner, such as a sipper bottle and bowl, to stimulate increased water intake. Some ferrets enjoy playing in their water, so a heavy crock/bowl is recommended. Actual water consumption can be significantly affected by the diet provided (dry versus moist) as well as other environmental factors. Supplements or additives should not be added to ferrets' water.

Rats

The Norway or brown rat (*Rattus norvegicus*) belongs to suborder Myomorpha (mouse-like rodents). Originating in temperate regions of central Asia, rats are nocturnal, burrowing, and colonial. Like most other myomorph rodents, rats are naturally omnivorous, eating grasses, seeds, grain, and occasionally invertebrates and can practice coprophagy based on diet. Rats are quiet, clean, and easy to maintain. They make excellent pets, especially if socialized from an early age.

Rats feed primarily at night and sleep during the day but can adapt to environmental conditions. They tend to avoid new foods therefore diet changes must be made gradually. Diet should be based on a commercial pelleted feed (rodent block) supplemented with a large variety of fresh and dried fruits and vegetables and smaller infrequent amounts of animal and insect-based proteins. provided in moderation. Seed-based diets should be avoided to prevent selective feeding, malnutrition, and obesity. Limited amounts of seeds, nuts, raisins, etc. should be provided only as treats and to encourage foraging behavior. Protein recommendations vary from 16-20%. Rats practice coprophagy, which provides nutrients such as B-vitamins and bacterial protein. Young rats are reared communally, and nursing responsibilities are shared. Weaning occurs at 21 days of age.

Like other rodents, rats are monophydont (have only one set of teeth) and only their incisors grow throughout their life. Their dental formula is 2(I 1/1, C0/0, P 0/0, M3/3). The ratio of upper incisor to lower incisor length is approximately 1:3, which inexperienced practitioners often mistake for overgrowth of the lower incisors.

Food is grasped in the forepaws, chipped by the incisors, and ground down by the molars. Only one side of the mouth is in apposition at a time during mastication. Cheek tissue can be drawn into the diastema, closing off the back of the mouth to allow gnawing without debris passing into the oral cavity.

The esophagus enters the stomach in the middle of the lesser curvature. They are monogastrics with the stomach divided into two parts by a limiting ridge. Situated at this junction, the limiting ridge prevents a rat from vomiting. Myomorphs such as the rat are unable to vomit also, in part, due to the crural sling (muscle) and the innervation of the diaphragm; therefore, pre-operative fasting is not required. The non-glandular forestomach is lined by squamous epithelium, while the distal glandular part is lined by fundic glands containing chief and parietal cells. Like other myomorphs, rats have a relatively large cecum and long colon. The cecum is not septate as in other species. GI transit time for the rat is reported to be between 8-16 hours depending on diet factors.

Mice

The domesticated mouse (*Mus musculus*) originated in temperate Asia. Mice belong to the suborder Myomorpha (mouse-like rodents) and are primarily nocturnal. They are social and prefer to live in small groups dominated by a single adult male. Mice are omnivorous in the wild, eating grasses, seed, grain and occasionally invertebrates. Mice naturally hoard food items, thus food consumption is hard to measure. Diet should be based on a commercial pelleted feed (rodent block) supplemented with a large diversity of fresh/dried fruits and vegetables and smaller infrequent amounts of animal and insect-based proteins. Seed-based diets should be avoided to prevent selective feeding, malnutrition, and obesity. Limited amounts of seeds, nuts, raisins, etc. should be provided only as treats and to encourage foraging behavior. The diet should contain 4-5% fat and 16% or more protein for maintenance; 7-11% fat and 20-24% protein for breeding. Fresh water should be available at all times (monitor water bottles for malfunction). Like all rodents, mice practice coprophagy, which provides nutrients produced via hindgut fermentation.

Mice have elodont incisors and anelodont check teeth. The mouse dental formula is 2(I 1/1, C0/0, P 0/0, M3/3), upper/lower incisor length ratio is 1:3 and incisors grow throughout their life. Mice are monogastric, but there is a division (the "limiting ridge") between the forestomach and the glandular stomach. The forestomach is lined by squamous epithelium and the glandular stomach by cardiac, fundic, and pyloric glands. Like other myomorphs, mice have a relatively large cecum and long colon. Because of the lower esophageal sphincter, the limiting ridge, crural sling, and innervation of the diaphragm, mice cannot vomit. The liver of the mouse is composed of four lobes: median (divided into left and right portions), left and right lobes, and caudate lobe. The life span of mice averages 1.5-2 years; up to 3 years is reported.

Gerbils

The Mongolian gerbil (*Meriones unguiculatus*) is a Myomorph rodent native to the desert regions of Mongolia and northeastern China. It is a social animal that is active during the day and night. Gerbils are adapted to a desert environment, require little water and produce only small amount of concentrated urine. Their feces are small, dry, and hard. For this reason, gerbils are nearly odorless and require less cleaning than do other rodents.

Their dental formula is 2(I 1/1, C0/0, P 0/0, M3/3). Wild gerbils have a variable, seasonal diet of leaves, seeds, and insects. If captive gerbils are often fed a seed-based diet, which is nutritionally incomplete. A pelleted rodent chow with 15-20% protein is recommended instead. For reproduction, dietary protein may need to exceed 20%.

Although wild gerbils can obtain almost all the water they need from available fruit and vegetable matter, commercial pellets do not contain adequate moisture to meet these needs. The captive diet should be supplemented with small amounts of fresh/dried fruit and/or vegetables such as apples and carrots, occasional seeds and insect-based nutrients as well as fresh, clean water available at all times. Fat in the diet should be 4% or less, as gerbils are prone to high cholesterol levels. Gerbils appear to have a higher need for magnesium than rats may develop alopecia and increased susceptibility to seizures when fed a Mg-deficient diet. Gerbils are not typically coprophagic when fed a nutritionally complete diet.

Hamsters

Hamsters are Myomorph rodents of Middle Eastern and Asian origin. A number of different species of hamster are kept as pets: the Syrian or golden hamster (*Mesocricetus auratus*); the closely-related dwarf hamsters—the Russian dwarf, Siberian or Djungarian hamster (*Phodopus sungorus*), the Campbell's Russian dwarf hamster (*Phodopus campbelli*), and the Roborovskii (Robo) or desert hamster (*Phodopus roborovskii*); and the Chinese or striped hamster (*Cricetulus griseus*). Hamsters are generally solitary and nocturnal, although dwarf hamsters may live in family groups.

Hamsters are nocturnal, burrowing animals that prefer a deep layer of bedding. The word hamster comes from the German word hamstern, which means to hoard. Hamsters naturally hoard food therefore the cage must be cleaned frequently to remove hoarded food and prevent ingestion of moldy food items. Hamsters are desert adapted however they are not heat tolerant; they burrow to escape heat. They will hibernate at low environmental temperatures, which may be mistaken for extreme debilitation or death. Hamsters live 1.5-3 years depending on species.

Hamsters are omnivorous and coprophagic. The feeding pattern of adults is typically a 5-minute feeding period, followed by a 2-hour fast. Diet in the wild includes grains, seeds, plants, and some invertebrates. Dietary protein requirements are lower for hamsters than for mice/rats. Pet hamsters are usually fed a commercial hamster mix containing 16-20% protein, 60-65% carbohydrate, and 5-7% fat, which can be supplemented with fruit, vegetables, nuts, and animal protein (e.g., egg, cheese, insects). Seed-based diets are common but lend towards selective feeding. Commercial pellets are less likely to result in nutritional deficiency, obesity, and diabetes. Hay is offered to increase fiber consumption, and sugary treats are kept to a minimum. Hamsters should have wood, branches, etc. to gnaw on for enrichment, and fresh water must be available at all times (water bottle malfunction is relatively common).

The hamster dental formula is 2(I1/1, C0/0, P0/0, M2-3/2-3). The four prominent incisors are elodont while the cheek teeth are anelodont. The crown of the mandibular incisors is longer than the crown of the maxillary incisors—commonly mistaken to be overgrown. The yellow color of the incisors is due to iron pigments. The cusps of cheek teeth retain food and thus hamsters are prone to dental caries formation (males > females).

Hamsters have large cheek pouches, invaginations of the oral mucosa used for transporting food, bedding material, and young. When swimming, hamsters can inflate their cheek pouches in order to increase buoyancy. Cheek pouches extend to the shoulders/scapulae and are retracted by the buccinator muscle, which evolved from the trapezius muscle. Hamsters empty their cheek pouches by pushing cheek contents forwards with the paws. When weighing a hamster, the weight of food in the cheek pouches must be taken into consideration.

Hamsters have a distinct non-glandular forestomach (*pars cardiaca*) that resembles the rumen. Like a rumen, it has high pH and contains microorganisms for fermentation. The glandular stomach (*pars pylorica*), is like that of a monogastric animal. The differential in pH between the two compartments indicates that food does not mix between them. The presence of VFAs (mainly acetic acid) indicates that some fermentation does go on here, although food only stays in the stomach for about 1 hour. The forestomach is lined by squamous epithelium while the glandular stomach region is lined by glandular mucus membranes. Stomach bacteria are mainly gram-positive with some gram-negative coliforms. The two chambers are separated by a muscular sphincter which may regulate passage of food between the two sections. The distal esophagus has a muscular sphincter and enters the stomach just cranial to this constriction. The lesser curvature of the stomach is greatly reduced such that the cardia and the pylorus are very close; therefore, vomiting is impossible for hamsters.

The cecum is sacculated, containing bacterial microflora including *Bacteroides* spp. and *Lactobacillus* spp. plus a few coliforms and clostridia. The pH in the cecum is higher than in the non-glandular stomach, suggesting more active fermentation. The liver has four lobes: right and left medial, and right and left lateral. Hamsters are permissive hibernators; low environmental temps stimulate food gathering, and at 41°F and below they enter deep sleep.

Degus

The degu (*Octodon degus*) is native to the western foothills of the Andes in northern and central Chile. It is the most populous mammal in Chile, where it is considered an agricultural pest. Degus are rodents belonging to the suborder Hystricognathi (porcupine-like rodents), based upon jaw musculature and skull structure. They belong to the family Octodontidae, and genus *Octodon*, which refers to the figure-eight shape of their cheek teeth. Its other names include brush-tailed rat and trumpet-tailed rat. Degus are semi-fossorial, diurnal, and do not hibernate.

Degus are adapted anatomically and behaviorally to a fibrous, low-carbohydrate diet. The wild diet includes a wide variety of plants, grasses, roots, seeds, fruit, and livestock droppings. Captive diet should consist of high-quality hay, limited high-fiber herbivore rodent pellets (e.g., chinchilla pellets), a small amount (20% of daily intake) of greens such as kale, collard, romaine lettuce and dandelion greens, broccoli, green beans, and roots such as beets, carrots, parsnips, sweet potatoes, and turnips. Degus produce the enzyme L-gulonolactone oxidase and therefore do not require dietary supplementation with vitamin C.

Improper diet is a major cause of disease in captive degus. To reduce the risk of diabetes and cataracts, foods containing sugar (e.g., apples, raisins, other fruits, breakfast cereal, honey treat sticks) should be avoided. Note that most pelleted feeds (including guinea pig chow) contain molasses, which is used as a binder. Starchy foods may also pose a risk.

Degus are similar anatomically and to both guinea pigs and chinchillas. They possess large, powerful incisors which are yellow to orange from iron pigmentation. The dental formula is 2(I 1/1, C 0/0, P 1/1, M 3/3). Both the incisors and the cheek teeth are elodont (aradicular).

The degu stomach is simple and glandular, and the intestine is the primary site of nutrient absorption. Degus possess a large, functioning, haustrated cecum where microbial fermentation of ingesta occurs. The colon lacks haustra, and colonic absorption of water from digesta results in very dry feces. GITT is very fast at only 5.1 hours. Degus are cecotrophic and coprophagic: one study showed that 38% of feces produced in 24 hr period were reingested; 87% of this activity occurred at night. Cecotrophy and coprophagy begin as early as 3 days of age.

Prairie Dogs

The black-tailed prairie dog (*Cynomys ludovicianus*) is a stout, short-legged ground squirrel native to the Great Plains of the United States. Prairie dogs belong to suborder Sciuromorpha (squirrel-like) rodents. They are herbivorous, having a functioning cecum. In their natural habitat, prairie dogs eat primarily grasses, prairie herbs and weeds, and occasionally roots and insects. Prairie dogs are diurnal and spend much of the day foraging for food; they do not store food in their burrows.

Captive prairie dogs should receive unlimited grass hay and a measured quantity of pelleted food; prairie dog pellets are preferred, but rodent block and rabbit pellets are also suitable. Alfalfa, fresh greens and vegetables can be offered as treats. Calorie consumption for the prairie dog varies throughout the year, increasing in the fall; caloric intake should be regulated to prevent obesity.

Fresh water should always be available, provided daily in a bowl or sipper bottle. Fluid consumption varies depending on diet. Wild prairie dogs can get most of the water they need from the plants they eat; however, dry pelleted diets necessitate more water intake than fresh foods do.

Prairie dogs are more difficult to care for than other pet rodents. As a native species, they are also illegal to keep in some states. Prairie dogs are relatively hard to handle and may bite. To syringe feed or medicate, handlers may need to use a burrito wrap or heavy leather gloves. Gavage feeding is complicated by the presence of a palatial ostium; with a 5-Fr nasogastric tube it is possible but unlikely to be very effective.

Prairie dogs are in the squirrel family. Their incisors are elodont (aradicular) whereas their cheek teeth are anelodont. The dental formula for the prairie dog is 2(I 1/1, C 0/0, PM 2/1, M 3/3). Incisor enamel is normally orange-yellow in color. Prairie dogs are hindgut fermenters that require adequate dietary roughage. They are also coprophagic. Normal droppings resemble those of guinea pigs, dry and oval in shape. Prairie dogs have a triad of scent glands surrounding the anus. Their white papillae resemble fly larvae when they are winked. These are not removed as an elective procedure in prairie dogs.

True hibernation does not occur in prairie dogs, but dormancy may occur for brief periods during cold weather. Dormancy can also be stimulated in captivity by sudden drops in temperature. Prairie dog pups appear above ground at approximately 5-6 weeks and are weaned at about 5-7 weeks. The majority of prairie dog pups in the pet trade are harvested each spring from the wild. Prairie dogs do not breed well in captivity.

Hedgehogs

The African pygmy hedgehog (*Atelerix albiventris*) is native to West and Central Africa, one of many hedgehog species found in Africa, Europe, Asia, and the Middle East. Hedgehogs are monogastric omnivores, consuming a variety of invertebrates, frogs, lizards, snakes, mice, eggs, fruit, and fungi. They are generally solitary and nocturnal. They inhabit a wide range of habitats including grassland, scrub, savannah, and suburban gardens. They live in brush piles, rock crevices, and burrows. They are able to climb, dig and swim. The African pygmy hedgehog can hibernate during periods of cool, dry weather, remaining torpid for periods of up to six weeks but emerging during warm intervals.

Hedgehogs exhibit a unique behavior known as self-anointing or anting, which can be triggered by any new or strange object or substance. The new object is licked until thick, frothy saliva accumulates. The saliva is then carefully placed onto the spines of both sides of the body with the tongue. This elaborate process may continue for up to 20 minutes. The behavior may serve as a sexual signal (confined to the breeding season in the wild) or protection against predators. Substances that cause self-anointing behavior are typically novel or irritating (e.g., hedgehogs have been observed to rub toads against their spines). This behavior has been observed in hedgehogs as young as 15 days of age.

Hedgehogs have 36 teeth. The dental formula is: 3/2 incisors, 1/1 canines, 3/2 premolars, and 3/3 molars. Hedgehogs have a simple stomach and no cecum. The GI transit time averages 12-16 hours.

African pygmy hedgehogs appear not to have a physiologic need to hibernate and will not do so if kept warm and well fed. They may enter hibernation if the environmental falls below 18 °C (65 °F). Metabolism, heart rate and respiratory rate will decrease, while synthesis of endogenous heparin increases, preventing thrombus formation. This decrease in metabolism renders hedgehogs more susceptible to infection, therefore, pet hedgehogs should be prevented from hibernating.

The natural diet of hedgehogs includes insects, worms, snails, slugs and occasionally small vertebrates and fruit. The nutritional requirements of hedgehogs are not known. They can be successfully maintained on dry low-fat cat food and commercial hedgehog diets are also available for use as the staple diet but quality of ingredients should always be evaluated. Supplement the diet with earthworms, mealworms, crickets, and a small amount of chopped fruits and vegetables. Hard-boiled or scrambled eggs, evaporated milk, cottage cheese, wax worms, pinky mice, and vegetable or meat baby foods provide additional variety to the diet. An insect or meat-only diet should be avoided in order to prevent metabolic bone disease.

Hedgehogs can learn to use a sipper-type watering system or water can be provided in a shallow pan. Provide fresh water daily. Offer food in the evening. Limit the amount to what an adult will consume by morning. Use heavy crocks and food bowls that cannot be tipped. Ad lib feeding, over-consumption, and inactivity commonly lead to obesity in African hedgehogs. The problem can become so severe that the patient can no longer fully roll up. Obesity can lead to other metabolic derangements and health problems, including hepatic lipidosis. Treat obesity by gradually reducing the quantity of food until the amount offered is consumed overnight. Limit foods high in fat: meal worms, wax worms, etc. Offer live insects in the cage to promote normal hunting and burrowing activity.

Critical Care Feeding of Exotic Companion Mammals

Micah Kohles, DVM, MPA

Session #2902

Affiliation: From University of Nebraska-Lincoln, Lincoln, NE, Oxbow Animal Health, Murdock, NE 68407, USA, Woodland Animal Hospital, Hickman, NE 68372, USA.

Introduction

Exotic companion mammals require diets specifically formulated for their nutritional needs. During times of anorexia or poor nutritional status, assisted feeding may be required and their requirement for specifically formulated diets does not change.[1-6] For many of these species, anorexia of even 12-24 hours in duration can have serious health implications and even lead to mortality, so providing vital nutrients when voluntary intake ceases is crucial. Powdered recovery diets mixed with water also provide essential hydration as well as routes to administer medication, in addition to nutritional support. Case-specific calculations of feeding amounts using animal body weight, appropriate metabolic factors, and activity and growth multipliers may help avoid over- or under-feeding. While duration of assisted feeding can vary, goals of recovery feeding should always be to return the animal to voluntarily consuming its standard diet.

What is Critical Care or Recovery Feeding?

Recovery or assisted feeding products should be formulated and designed with the intent of hand feeding animals unwilling or unable to eat on their own. Calorically dense, nutritionally complete powdered products reduce volume intake while still providing key nutrients for recovery and health. Because these products are meant to be mixed with water, they also supply vital hydration to animals in sub-optimal nutritional and dehydration status. When selecting a recovery or assisted feeding formula, it is vital to use a formula that aligns with the animal's nutrient needs. When targeting exotic companion mammals (ECM), products designed for dogs and cats may not be appropriate. It is also important to select products formulated with ingredients that appropriately mimic the normal diet to encourage voluntary feed intake. Biologically relevant ingredients are easily digested by the animal and enhance palatability to stimulate appetite appropriately during recovery.

General ingredient recommendations specific to ECM nutritional needs are as follows:

- Herbivores: grass-hay-based formulas with added plant-based sources of protein, fat as well as omega fatty acids, prebiotics, and amino acids for recovery; natural flavors for appetite stimulation.
- Omnivores: balance of plant, animal, and insect ingredients that are biologically relevant and appropriate for these species.
- Carnivores: animal-protein-based with high caloric density and balanced amino and omega fatty acids profiles; plant-based protein isolates should be avoided.

Artificial colors, preservatives and flavors have no nutritional benefit and should be avoided.

When to Assist Feed

Assisted feeding in ECM is appropriate, or even vital, in a wide variety of situations and should occur anytime an animal becomes anorexic. Anorexia is a non-specific clinical sign and investigation into root causes should begin immediately. For certain ECM species, especially hind gut fermenters such as rabbits, guinea pigs and chinchillas, anorexia bouts lasting even 18-24 hours constitute a potential emergency. Assisted feeding can and should occur any time an animal is anorexic or has potential to go off feed including, but not limited to, use as or during times of the following:

- Appetite stimulant
- Malnutrition
- Stress
- Support during specific disease states
    - Gastrointestinal stasis
    - Dental disease
    - Anemia
    - Pancreatitis
    - Insulinoma
    - Metabolic disease
    - Cancer support
- Severe weight loss
- Geriatric patients
- Medication carrier
- Pre-surgery and post-surgery
- Receiving protocols

While assisted feeding in the clinical setting may be best, it is not always practical for long-term patients. Teaching owners of pets assisted feed or having handouts and videos to send home can be a great tool to ensure the animals best chance of recovery. This can also be made easier by having assisted feeding products for sale in the clinic or on hand to send with patients. Stress alone can trigger anorexia in many ECM species, so pets entering a new home, changing diet, and otherwise entering a period of stress may also benefit from 1-2 days of assisted feeding.

How Much to Feed?

Most companies that make critical care feeding products will have charts or websites with instructions on how to calculate feeding amounts. Charts often are oversimplified for convenience sake because they are typically used as a quick-reference guide in clinical situations. The most accurate way to calculate feeding doses is via species- and situation-specific calculations. One example of an available reference is Oxbow's Vet Connect website (https://www.oxbowanimalhealth.com/vet-connect), which utilizes specific formulas to calculate individual feeding recommendations. They can be used with even greater detail specific to the animal in treatment.

The first step in calculating an animal's energy requirement is determining metabolic body weight (MBW; $BW_{kg}^{0.75}$). Using Kleiber's law, this equation states, for the vast majority of animals, the metabolic rate scales to the 3⁄4 power of the animal's mass (body weight in kilograms). Using MBW, the resting energy requirement (RER) can be calculated by multiplying by a factor specific to species or groups of species: MBW × factor. These factors are well defined for dogs and cats in the literature, but for ECM and other exotic species, field metabolic rates have much more recently been investigated and validated.[7]

Metabolic energy requirements (MER) are calculated by multiplying RER by an appropriate activity or growth multipliers: RER × multiplier. Research has been conducted by Omaha's Henry Doorly Zoo, that determined safe multipliers for exotic mammals and reptiles at 1-2 and 1-3 for exotic birds. Other multipliers may be used for various physiologic states or metabolic goals and include the following:

- Adults at maintenance = 1.0
- Young/growing = 1.5
- Weight gain = 1.2-1.8
    - Use *ideal* MBW
- Weight loss = 0.8-1.0
    - Use *ideal* MBW
- Seniors = adjust for weigh loss or gain as necessary

Once you have calculated the MER (kcal), product feeding amounts can be calculated. Grams of product to feed are calculated by dividing MER by kilocalories per gram of product: MER/kcal per gram.

- Kcal/gram (Oxbow Critical Care)
    - Herbivore: 2.67
    - Omnivore: 3.6
    - Carnivore: 6

If you prefer to calculate feeding recommendations on a volumetric basis, dividing feeding grams by grams per tablespoon of dry product can be calculated: MER ÷ kcal per gram ÷ grams per tablespoon.

- Grams/tablespoon (Oxbow Critical Care)
    - Herbivore: 9
    - Omnivore: 8
    - Carnivore: 4

Example

A 0.5-kg, 3-month-old guinea pig eating only Oxbow's Critical Care

MBW: $0.5^{0.75} = 0.59$

RER: $0.59 \times 140 = 83.24$

MER: $83.24 \times 1.5 = 124.87$ kcal

Grams Critical Care feed: 124.87 kcal ÷ 2.67 kcal per gram = 46.78 grams Critical Care per day

Tablespoons Critical Care feed: 46.78 grams Critical Care ÷ 9 grams per tablespoon = 5.20 tablespoons Critical Care per day

How Long to Assist Feed?

Duration of assisted feeding is extremely case-specific. The most generic answer would be to continue assisted feeding until the animal has returned to voluntary dietary intake. During this time, investigation into root causes should be carried out to determine if further medical diagnostics or treatment is necessary. If after 3-5 days patient status has not improved, underlying issues are likely preventing full recovery. While most Critical Care and assisted feeding formulas are designed to be short-term solutions, there are cases in which a recovery formula can be fed on a longer-term basis under close supervision. However, goals of recovery feeding should always be to return the animal to voluntarily consuming its standard diet.

What About Hydration?

Chronic and acute dehydration are constant threats for ECM species. Understanding the role hydration plays in animal well-being and recovery is essential to properly support ill and diseased animals. Independent of the route of hydration support, monitoring patient hydration status and ensuring appropriate levels of hydration support are key.

Rabbits

- Typical water intake is about 50-150 ml/kg (22-46 ml/lb) daily
- 2.5-kg rabbit on a dry diet will take approximately 160 ml of water, or 65 ml/kg, daily
    - This amount is approximately twice the intake of an average human adult derived from both food and drink
- Summary: 50-100 ml water/kg body weight
    - Range: 50-150

Guinea pigs

- 21.7 ml/100 g BW/day in 6-week-old male guinea pigs (312 ± 13 g) that were individually housed on sawdust[8]
- 7.5 ml/100 g BW/day in male guinea pigs (698 ± 19 g) fed a non-purified diet containing 20 percent crude protein[9]
- Summary: 75-125 ml water/kg body weight
  - Range: 75-215

Hamsters and gerbils

- Male and female golden hamsters consume, on average, 8.5 ml water/100 g BW/day
  - Males consumed 5 ml water/100 g BW/day, while females consumed 14 ml water/100 g BW/day[10]
- Chinese hamsters water intake 11.4 ml/100 g BW/day for males and 12.9 ml/100 g BW/day for females
- Water intake for golden hamsters was found to be 4.5 ml/100 g BW/day in males and 13.6 ml/100 g BW/day for females
- Summary: 75-125 ml water/kg body weight
  - Range: 45-140

Acknowledgments: I thank Cayla J. Iske, PhD for her contributions to this manuscript.

References

1. Lichtenberger M, Lennox AM. Exotic companion mammal emergency medicine and critical care. *Proc Annu Conf Assoc Exot Mamm Vet.* 2009.

2. Paul-Murphy J. Critical care of the rabbit. *Vet Clin North Am Exot Anim Pract.* 2007;10(2):437-461.

3. Fisher PG. Standards of care in the 21st century: the rabbit. *J Exot Pet Med.* 2010;19(1):22-35.

4. Graham JE, Doss G, Beaufrère H, eds. *Exotic Animal Emergency and Critical Care Medicine.* John Wiley & Sons; 2021.

5. Hawkins MG, Graham JE. Emergency and critical care of rodents. *Vet Clin North Am Exot Anim Pract.* 2007;10(2):501-531.

6. de Matos R, Morrisey JK. Emergency and critical care of small psittacines and passerines. *Sem Avian Exot Pet Med.* 2005;14(2):90-105.

7. Nagy KA, Girard IA, Brown TK. Energetics of free-ranging mammals, reptiles, and birds. *Annu Rev Nutr.* 1999;19(1):247-277.

8. Liu CT. Energy balance and growth rate of outbred and inbred male guinea pigs. *Am J Vet Res* 1988;49: 1752-1756.

9. Tsao CS, Young M. Effect of dietary ascorbic acid on levels of serum mineral nutrients in guinea pigs. *Int J Vitam Nutr Res.* 1989;59:72-76.

10. Fitts DA, St. Dennis C. Ethanol and dextrose preferences in hamsters. *J Stud Alcohol.* 1981;42(11):901-907.

How to Create Mutual Respect and Mutual Trust in the Examination Room

Jan Hooimeijer, DVM, RPBC

Session #2573

Affiliation: From Consultancy Practice for Birds, Schoolstraat 7 A 7941 CA Meppel, The Netherlands.

Abstract: Addressing welfare and wellbeing of parrots is our responsibility as avian veterinarians. It all starts in the examination room by acting as a positive role model making a difference, creating mutual respect and trust with the bird. It is an exciting learning experience for the parrot and for the caretaker. Showing love can be considered as the opposite of showing respect.

Showing respect is acknowledging our body language, the cognitive abilities and the intelligence of parrots as prey animals. Ignoring the negative effects of human body language and human attitude is not showing respect. Acknowledging the need for privacy is another way to make the parrot experience that we show respect.

Within the Consultancy Practice for Birds, the 5-step protocol is developed as a tool to create mutual respect and mutual trust, reducing the stress of novel experiences including handling. Understanding parrot behavior and behavioral problems is crucial to be able to show respect for the birds.

Definitions are crucial to describe and to understand parrot behavior. Within the Consultancy Practice for Birds, definitions are formulated concerning normal behavior, undesired behavior, desired behavior and enforced behavior. The definition of displacement behavior is formulated by Niko Tinbergen. Using displacement behavior within the examination room is described as a method to work with parrots showing behavior problems based on insecurity.

Parrots are, in general, kept under extreme unnatural circumstances in captivity. Wing clipping can make it possible to offer parrots more freedom as companion birds and to become part of the family flock inside and especially benefit outdoors from the social interaction, exercise, sun and fresh air. The disadvantages of having a parrot on the shoulder are discussed.

Introduction

The major dogma within veterinary science is to do no harm. Working with animals as veterinarians, it is essential to know and understand the basics of handling and manipulating the animal in a professional manner. A professional manner is inclusive of avoiding unnecessary stress as well as avoiding trauma to the animal, the handlers, and caretakers. Reducing stress is an important responsibility for anyone caring for animals.

In general, there is still a lack of awareness that parrots are non-domesticated prey animals being kept in captivity under unnatural and often bird unfriendly circumstances.[1] In general, parrots are still not acknowledged as highly intelligent animals with remarkable cognitive abilities. This is the reason for a wide variety of behavioral problems depending upon the individual parrot, species, and circumstances surrounding the bird.

In the context of the veterinary clinic we have to realize that owners may be nervous and uncertain about what to expect during the consultation, examination and treatment. We need to consider that the attitude and insecurity of the owner can have a negative influence on the behavior and the welfare of the parrot. In general, owners do not realize that their presence can create insecure behavior, and not the presence of a knowledgeable avian veterinarian. In my examination room I always compare the behavior of a child in the presence of the nursery school teacher with the behavior of the same child at home in the presence of the parents.

We often have to discuss basic behavioral issues with the caretaker before approaching and dealing with the parrot. It is important that owners demonstrate happy and positive behavior themselves to prevent insecure behavior in the parrot.

The caregiver is asked not to look at the parrot under any circumstance, especially when the parrot is showing insecure behavior. The owners are to look at the whole exam room, and to me, instead.

When the parrot shows fear, every time the caregiver is giving attention (including simply looking at the bird), we have to make the caretaker aware of the rewarding effect of any response that is confirming the fear and uncertainty. For that reason, caretakers are not allowed to reassure the parrot because it is in general not productive, especially when owners are showing that they are not comfortable, themselves, with the situation. This reassuring is rewarding the parrot's fear and creates a stressful situation.

The first impression of the veterinarian towards the parrot is very important. It is the first opportunity to make the parrot experience whether the veterinarian is showing respect for the bird or not.

It is crucial to start with showing to be happy and positive and not paying any attention to the parrot entering the examination room and during the conversation with the caregiver. It means that the bird is not put on the table in the middle of the attention but put aside in its cage or on a perch or branch. It is providing the parrot the opportunity to observe and experience that the veterinarian is the opposite of intimidating and feels comfortable in the presence of the parrot as if the bird is not there. To me this is the first step to make the bird experience that I show respect for the bird.

Behavior

Behavior in nature is based on innate behavior and founded on learning, experience, intelligence and cognitive abilities. Behavior is driven by the urge to survive as an individual and as a species.

Working with parrots in captivity we need to understand that the behavior of parrots is determined by different factors including hereditary factors, as well as being determined by living in captivity under extremely unnatural circumstances.[1] To understand behavior we have to learn about the principles of ethology developed by the founders of ethology and Nobel prize winners Niko Tinbergen, Konrad Lorenz, and Karl von Frish. A well-known ambassador of ethology is Jane Goodall who has shown that showing respect for chimpanzees in nature is rewarded by a relationship based on mutual respect and mutual trust.

Niko Tinbergen has formulated the four questions below as a tool to understand and interpret behavior.

1. What is the function of the behavior?
2. What is causing the behavior?
3. How to explain the behavior within the context of ontogeny?
4. What is the evolutionary origin of the behavior?

Looking for differences between showing love and showing respect, after answering the four questions, it is obvious that showing love is behavior that is driven by the need for reproduction to ensure the future of the species. Hormones do play a crucial role within a mutual love relationship meant to produce offspring. Relationships based on mutual respect and trust are not based on hormones and not driven by the need to create offspring. Creating a relationship with a parrot based on love can be considered as unnatural and must be very confusing for the bird. Creating a relationship based on mutual respect and mutual trust results in a long term relationship, ensuring the welfare of the bird. Relationships based on mutual respect and mutual trust are crossing the borders between species as Jane Goodall has shown. The relationships between humans and non-human domesticated animals are crossing the lines between species all the time. The same for many examples of relationships between individuals of different species.

Parrots are built and behave as prey animals

Parrots are prey animals in nature. The fear of being killed as a prey animal determines a major part of normal behavior in nature. Fear of getting killed as prey animals can also be expected as normal behavior within captivity.

A typical anatomic feature of parrots as prey animals is the positioning of their eyes. The eyes of parrots are positioned at the side of their head enabling parrots to observe the whole environment. The eyes of predators are positioned in a way enabling the animals to look ahead, using binocular vision. The eyes of humans are positioned as in predators like dogs, cats, owls and birds of prey having binocular vision. The perception of a parrot looking at humans that are talking and smiling to the parrot, showing their teeth, can be considered as being intimidated by a predator. Looking through the eyes of parrots we have to consider that our body language and attitude determines in different ways the body language and behavior of parrots. Working with parrots without understanding the consequences of the specific characteristics of parrots as prey animals is not showing respect for the birds and is a reason for parrots to develop insecure or defensive behavior. This insecurity is cause for a wide range of behavior problems.

Parrots need privacy

In nature, parrots are hiding in the foliage of the forest canopy. Other bird species exhibit the same behavior to avoid being noticed by predators. This explains the captive parrot's natural need to search for places to hide and their need for privacy. At the same time we know that the need for privacy is not restricted to prey animals. Lack of privacy can be considered as affecting the welfare of animals including humans.[2] Approaching a parrot within the context of the examination room we need to acknowledge that invading their territory, especially while sitting in their cage, can be a reason to create insecure behavior and therefore behavior problems.

Parrots are intelligent birds with amazing cognitive abilities

Learning more about the intelligence and cognitive abilities of parrots we can conclude that parrots do understand our posture and attitude. Parrots learn by observing other birds and other animals including humans as part of their environment. Parrots draw their conclusions and behave accordingly. Dealing with parrots is like dealing with children that are eager to show and demonstrate what they have learned. The model/rival technique, described and used by Pepperberg is acknowledging the intelligence and observational learning skills of parrots.[2]

Allowing myself as an author some anthropomorphism, from the viewpoint of parrots, it must be frustrating to experience not being appreciated for their intelligence and talents but to be considered as just beautiful, funny, and cute.

Understanding behavior is crucial to be able to show respect. It is important to educate caretakers before working with the bird to allow them to observe, in an educated way, the proper approach, and to learn from the interaction between the bird and the knowledgeable veterinary technician and avian veterinarian.

Definitions of Behavior

Normal behavior

Normal behavior is species and individual-specific behavior that is not meant to manipulate the caretaker. Examples include: Eating, drinking, grooming, playing, sitting in the cage, sitting on the hand. It is the experience of the author that the outcome of any response is rewarding normal behavior and the bird will become insecure and develop behavior to manipulate the caretaker to get attention. The behavior that started as normal behavior can easily become an undesired behavior meant to provoke a response.

Desired behavior

Desired behavior is created or manipulated by the caretaker in a way that the animal does not experience that the caretaker wants the animal to perform that specific behavior. It can be compared with manipulating the behavior of children with autism, knowing that those children are very sensitive and will respond in a negative way when they experience a command and even a friendly request. Rewarding the desired behavior, including rewarding their intelligence, is showing respect for the bird and will create self-esteem, mutual respect and prevent undesired behavior.[3]

Undesired behavior

Undesired behavior is meant to manipulate the behavior of the caretaker. Undesired behavior is meant to elicit a positive or negative response. In general, undesired behavior is meant to create a problem. This may include screaming, biting, feather picking, talking, and whistling to get attention. The outcome of rewarding undesired behavior is obvious. Any response like showing fear, frustration, anger, concerns and laughter can be considered as a reward and reason to develop even more undesired behavior.

Enforced behavior

Enforced behavior is the result of a command, any kind of force, or behavior, out of fear for punishment. Even a request can be considered as a command. In general, behavior that is motivated to get a reward can be considered as enforced behavior, making it the opposite of desired behavior.

Displacement behavior

Tinbergen[4,5] as one of the founders of ethology has described displacement behavior in gulls showing territorial behavior. Instead of a fight or flight response, the gulls started to preen themselves, started to pull at grass as if they were collecting materials to improve the nest and looked away from other gulls. In a conflict situation in nature (as well as in captivity) parrots have the choice between a fight response or a flight response. Displacement behavior can be considered as the ultimate alternative behavior when a fight or flight response are not the best choices, or are not possible. Displacement behavior seems to be completely out-of-order but is having the effect that the other parrot or the human caretaker becomes confused and overwhelmed, creating a new situation. Typical examples of displacement behavior are nail biting, scratching, grooming behavior, flapping the wings, shaking the tail, shaking the head, or looking at the sky with one eye.

Grooming and preening have been associated with normal behavior that birds perform when they are comfortable. Within the context of a conflict situation the same behavior can be shown as displacement behavior.[6] When we are responding to this behavior in a way that rewards displacement behavior, we can expect that this normal behavior in a conflict situation becomes undesired behavior to shape a response.

Working with parrots we can expect to see displacement behavior in different conflict situations. The most common conflict situation is when parrots as prey animals are approached by humans in an intimidating manner. We can also observe displacement behavior of parrots that are inside or on top of their cage demonstrating territorial behavior. Under those circumstances it is important to know how an avian species analyzes and combines its perceptual experiences to understand what is happening in the surrounding world.[7]

There are reasons to hypothesize that parrots utilize extensive information processing, that is at least to some degree, they do not simply react mindlessly to environmental stimuli. They actually process the stimuli and choose to react in a certain way.[8]

When displacement behavior is not recognized or not understood and caretakers are responding to displacement behavior in a way that is considered as rewarding, displacement behavior can easily result in abnormal repetitive behavior without the primary trigger that was causing the conflict situation to begin with.[9] It can be one of the explanations when we are dealing with complex behavioral problems like feather destructive behavior. It can be assumed that the type of displacement behavior has a genetic component and that the type of displacement may differ between species accordingly. It can also help explain why some species are more prone to feather destructive behavior.[10] It can be postulated that the way displacement behavior is expressed also will be influenced by environmental factors and observational learning ability of the parrots. Because of that, it has been argued that hand feeding baby parrots is contraindicated because birds need to learn by observing normal behavior of their parents, including normal displacement behavior in conflict situations.[11]

How to deal with displacement behavior

First, it is important to recognize and acknowledge displacement behavior and to consider that the parrot is demonstrating this behavior as they perceive a conflict. Any direct response to the bird at that moment can be considered a reward and can create maladaptive behavior. By showing displacement behavior as the caretaker or veterinarian, we are interrupting the conflict situation.

5-Step Dutch Behavioral Protocol Working With and Manipulating a Parrot

The most important aspect of dealing with parrots is the attitude of the person or persons caring for the parrot. A basic protocol for veterinarians and caregivers of parrots, creating mutual respect and mutual trust, can be summarized in five steps. This protocol can also be used with any parrot exhibiting a behavior problem.

Step 1

Make the bird observe without being the center of attention. Acknowledge the fact that parrots are prey animals by presenting a non-intimidating body language and posture. This way the parrot will observe that you are not a threat. It is important to show that you are comfortable in the presence of the parrot. This improves your perception as a positive role model. Show it all without any interaction with the parrot; don't even look at the bird. Ensure that you will have the bird's complete attention when you are doing your stuff and be silly, playing with a toy, throwing a little ball against the ceiling or whatever. Parrots feel comfortable with people who are comfortable with parrots and with themselves. Your behavior around the parrot will tell them all about you. This means that people often have to act and play a role around the parrot. Your behavior is demonstrating that you respect the parrot, resulting in mutual respect towards the veterinarian. What we create by doing so, at step 1, can be considered as desired behavior.

Step 2

The outcome of step 1 is creating a reason to tell, at step 2, the parrots that it's the most beautiful creature on this planet and to show how precious the bird is to you. Of course always use a non-dominating posture. The reward can be considered as reinforcing the behavior that has been created and manipulated by our attitude at step 1.

Step 3

Establish yourself as the best teacher the parrot can wish for. Show respect for his intelligence by telling yourself, in detail, what there is to see around you (i.e., birds in the trees, toys on the table, colors, shapes, sizes, objects). You also can describe the color of the tail and the beak. You can also describe your own body parts or your shoes. Parrots truly enjoy observing and learning. Acknowledging the intelligence is rewarding the bird for his intelligence, again in a non-intimidating way.

Step 4

After acknowledging/rewarding the intelligence, allow the bird to touch your pen, toy, paper or towel and praise the bird enthusiastically. Only let them touch items when you tell the bird, "now you may touch." Then they are in general very gentle when touching objects with their tongue. Ask the bird to participate in the learning process by offering him new, small, non threatening objects. Tell the bird that he/she may touch and feel the objects and allow it to investigate. Praise each positive move. At step 4 you also can present your hand and tell the bird that he/she may step up and praise the bird for doing so. This is the opposite of giving a command or making the bird step up motivated by a treat.

Step 5

Show the parrot that accepting new situations is okay because there are reasons to trust you under the circumstances that you have created. When the parrot shows fear, do not reassure the bird. Reassurance can easily create more fear and insecurity. It is the experience of the author that after steps 1 through 4 that mutual respect and mutual trust have been established. When the parrot fears a place or object like the towel, be understanding and start over again with step 1 of the protocol and follow through. Soon she/he will experience that there is no reason for fear. Then tell the parrot how smart and wonderful he/she is. This will help to reinforce the self-esteem that all parrots need to be able to develop normal behaviour. The outcome is to create a situation in which a parrot can accept novel situations. It can be a physical examination, taking a blood sample, grooming, wing clipping, gently towelling, or other situations.[3,12]

Dealing with Behavior Problems of Parrots

Dealing with behavior problems is looking for a diagnosis.[11] It is important to consider the circumstances and to understand the reason for the behavior and its consequences. In general, undesired behavior of parrots in captivity can be considered as behavior that started as a normal behavior, but under unnatural circumstances. Dealing with undesired behavior we have to consider that undesired behavior will become a greater problem when the consequence of that behavior is experienced by the parrot as a reward. Any response, (whether positive, negative, emotional, and even subconscious) to the undesired behavior can be considered a reward for the undesired behavior. Even ignoring the undesired behavior may be perceived as a reward.

Preventing undesired behavior

Prevention of undesired behavior requires understanding the reason for that behavior, including examining the circumstances and consequences of said behavior, as well as the natural behavior and body language of parrots as a social, intelligent prey animal. Prevention of undesired behavior involves ensuring their health and welfare by providing balanced and species-appropriate certified organic nutrition, sunlight (or appropriate artificial lighting), fresh air and exercise. Prevention of undesired behavior is encouraged by creating an enriched environment in which the parrot is allowed to express its intelligence and skills; providing toys and providing food in a way that it is stimulating parrots to express normal foraging behavior. Preventing undesired behavior is enabled with social interactions with other birds, other animals and humans. Taking a bird outside for a walk, a bike ride, a picnic or a family visit will help to prevent undesired behavior. Preventing undesired behavior also includes the understanding that the cage of a bird can be considered by the birds as their nesting site. In nature, birds spend their days and nights at the same location when they are breeding as part of their reproductive cycle. When at the nesting site, the bird may be defensive of intruders and express the need to demonstrate territorial behavior. Parrots benefit from a more natural day-night routine in which the bird gets 10-12 hours of sleep in a special small sleeping cage that is located in another room of the house, away from the location the bird spends the daylight hours. The understanding is that birds in nature do not sleep at night where they spend their daytime hours, foraging or drinking. When interacting with the bird, sitting on the shoulder is not encouraged as this may reinforce situational insecurity.

Redirecting undesired behavior by demonstrating displacement behavior

We have to consider that any response to undesired behavior can be perceived as a reward. Instead of responding to or ignoring the behavior, the advice is to act in a way that there is no problem with the demonstration of displacement behavior. Part of the displacement is "looking away" and demonstrating behavior showing no problem with the undesired behavior.

The result of demonstrating displacement behavior is that insecurity and aggression disappears. Over the years, the author has observed this behavior both as a birdwatcher in nature and as an avian veterinarian working with parrots in captivity. In a situation of undesired behavior, step 1 of the 5-step behaviour protocol, can be considered as showing displacement behavior, creating desired behavior that can be rewarded as step 2. Implementing the protocol in this order, undesired behavior is not rewarded. Having the bird watch us when we are not looking at the bird, acknowledging and rewarding their intelligence and rewarding the behavior with touching, feeling and even biting a specific object can all be done within 30 seconds. Using this protocol, undesired behavior is redirected into desired behavior and the bird is rewarded for that and immediately rewarded for their intelligence.

Wing Clipping

In the wild, flying is essential for the survival of parrots. Flying enables parrots to find food, water, and a safe place to sleep. Flying is a way to escape from predators; flying in flocks also offers protection from predators. The well known exception is the kakapo (*Strigops habroptilus*), which lives in a habitat that rendered flight expendable. For parrots in captivity that are kept as companion birds, flying is no longer essential for their survival. Within the Consultation Practice for Birds, wing clipping can be recommended for safety reasons but is mainly recommended because it makes it possible to provide the parrot more freedom and social interactions both inside the house and during the daily outside activities with the caretaker. Wing clipping may give the bird more freedom to become more a part of the family (flock). In the experience of the author, wing clipping done with a parrot-friendly attitude helps to prevent behavior problems and even prevents parrots from ending up in parrot rescue facilities.

Hand-Rearing Baby Parrots

The need to hand rear baby parrots in the past was the result of management failures concerning housing, nutrition and care complemented with a lack of knowledge about the natural breeding behavior of parrots. Over the years, hand-rearing baby parrots became a commercial activity because it encouraged females to produce more eggs.

The Dutch Parrot Foundation, supported by the Consultancy Practice for Birds has been successfully campaigning in the Netherlands against the hand-rearing of baby parrots and the trade in unweaned baby parrots. Since July 1, 2014, it is no longer allowed to separate a baby parrot from the parents until the baby parrot is able to eat completely on its own. This legislation has included parrots on the existing list with dogs, cats, rabbits, chimpanzees and other species where separating juveniles from their parents has been regulated for many years. This new regulation is based on data that has become available during the past 15 years demonstrating the negative consequences of separating baby parrots from their parents. A range of behavioral problems and welfare issues have been recognized as consequences of this practice. As described in other species, separating juveniles before or during the first imprinting phase can lead to irreversible behavior problems.

Some of these behavioral problems can occur after months or even years, especially when the birds become sexually mature. Insecure behavior, phobic behavior, biting behavior, feather picking and self-mutilation can become serious consequences of separating a baby parrot from the parents.[13-15]

Conclusions

We have to learn from parrots what they have known for tens of millions of years, by looking at the world through their eyes. Parrots-are among the most charismatic, intelligent, and social animals on the planet with a considerable life expectancy when we are doing a good job of caring for them. Consider and acknowledge parrots as prey animals. By adjusting our body language instead of showing an intimidating posture, creating insecure and defensive behavior, we show respect for parrots. Showing respect for their social skills, intelligence and cognitive abilities is another crucial part of making the birds experience our respect. Their natural learning process is by observing the behavior and skills of their parents, other members of the flock and humans in captivity. Never demonstrate insecurity or fear to parrots, as this will make the bird insecure and create behavior problems. Parrots are very perceptive of our insecurities. Using the 5-step behavior protocol creates a relationship based on mutual trust and respect.

It is vital to reward desired behavior. It is vital to reward the intelligence and acknowledge the cognitive abilities in parrots. It is vital not to reward normal behavior. It is vital not to create and not to reward undesired behavior. It is important to recognize displacement behavior and to use displacement behavior when we are in a conflict situation.

Taking care of the health, welfare, well-being and conservation of parrots in the world is our primary responsibility as avian veterinarians. The main focus should be on preventing health/behavior problems and to ensure their welfare in captivity. Parrots are honest and show whether you are doing a good job, just as children show whether their teacher is doing a good job. In that way parrots have become my best teachers because they correct me when I am wrong. I am grateful for the parrots in my life because they still teach me how to become a better avian veterinarian.

If Not Us, Who? If Not Now, When?

Acknowledgments: I thank Harrison's Bird Foods, The Netherlands for sponsoring the parrot events in the Netherlands. I thank the staff of the Consultancy Practice for Birds for their support and understanding. Last but not least I thank the owner of Roxy and Charlie for the support during the past years. Without this support, nothing could have been accomplished over the years.

References

1. Van Sant F. Seeing the rain forest through the trees. *Proc Annu Conf Assoc Avian Vet.* 1993;228-232.

2. Pepperberg IM, Hooimeijer J. Understanding and appreciating avian cognition to address behavioral issues. *Proc Annu Conf Assoc Avian Vet.* 2019;239-247.

3. Hooimeijer J. Dealing with behavior problems of parrots. Proc DVG. 2008.

4. Tinbergen N. *Spieden en Speuren in de Vrije Natuur.* Amsterdam, The Netherlands: Uitgeverij Ploegsma; 1959.

5. Tinbergen N. *Curious Naturalists.* New York, NY: Doubleday; 1958.

6. Sprijt BM, van Hooff JARAM, Gispen WH. Ethology and neurobiology of grooming behavior. *Physiol Tev.* 1992;72:825.

7. Zucca P. Mind of the avian patient: cognition and welfare. *Proc 8th Conf Europ Assoc Avian Vet.* 2007; 357-365.

8. Pepperberg IM. The African grey: how cognitive processing might affect allospecific vocal learning. In: Balda RP, Pepperberg IM, Kamil AC, eds. *Animal Cognition in Nature.* San Diego, CA: Academic Press; 1998.

9. Garner JP. Perseveration and stereotype-systems-level insights from clinical psychology. In: Rushen J, Mason G, eds. *Stereotypic Animal Behavior: Fundamentals and Applications to Welfare.* Wallingford, England, UK: CABI; 2006:121-152.

10. Seibert LM. Feather-picking disorder in pet birds. In: Luescher AU, ed. *Manual of Parrot Behavior.* Oxford, England: Blackwell Publishing; 2006:255-265.

11. Orosz S. Diagnostic workup of suspected behavioral problems. In: Luescher AU, ed. *Manual of Parrot Behavior.* Oxford, England: Blackwell Publishing; 2006:233-245.

12. Hooimeijer J. A Practical behavioural protocol for dealing with parrots. *Proc Annu Conf Assoc Avian Vet.* 2003;177-181.

13. Engebretson M. The welfare and suitability of parrots as companion animals: a review. *Anim Welf.* 2006;15:263-276.

14. Schmidt R. The influence of the breeding method on the behaviour of adult African grey parrots. Inaugural Dissertation zur Erlangung der Doktorwurde der Vetsuisse-Falkultat der Universitat Bern. 2004.

15. Hooimeijer J. Becoming a role model in the examination room handling parrots. *Proc Annu Conf Assoc Avian Vet.* 2005:97-109.

Exotic Companion Small Mammal Enrichment

Stefanie Kotschwar, RVT, BScH (Zoology, Nutrition), MSc (Zoo Conservation), AAS (Vet Tech), Diploma (Web Design and Equine Science)

Session #2610

Affiliation: From Lynwood Animal Hospital, 107-30 Stafford Rd, Ottawa, ON, K2H 1W8, Canada.

Abstract: There has been increased discussion in the animal world of providing enrichment to all animal species regardless of whether they are privately owned or are part of a zoological institution. Enrichment is important to our pet exotic small mammals to enhance their quality of care. The end goal is to provide an environment where species can express a whole range of species-appropriate behaviors and avoid any mental frustration or boredom. There are some minor pitfalls to providing enrichment such as time, money, accidents; however, these are far outweighed by all the positive benefits provided such as decreased anxiety, increased learning ability, and health monitoring. Enrichment can be divided into two general types: social and non-social enrichment. Social enrichment includes interaction with conspecifics (if they are a species found naturally in social groups) or with the owner. Non-social enrichment includes toys or materials made available to animals, as well as the actual enclosure and its location. While many general enrichment concepts are reviewed, there are some species-specific ideas that also are presented. While there are many exotic small mammal species, the focus of this presentation is on species frequently seen in veterinary clinics, such as rabbits (*Oryctolagus cuniculus*), guinea pigs (*Cavia porcellus*) and other small rodents.

Introduction

We know that providing enrichment in our exotic small mammal's life will positively affect their welfare, but have you really thought about what it means or how you can implement it? Do you know how to guide your client's into providing environmental enrichment for their own pets? There are many factors in deciding what kind of enrichment to provide to different species and how to assess whether it is providing a benefit or if adjustments need to be made. While it is easier to provide environmental enrichment at home, we can certainly provide it to our exotic small mammal patients during their short- or long-term visit to the clinic.

Not only does environmental enrichment provide a multitude of benefits to our exotic small mammals, but we as human caretakers also experience some benefits from these enrichment programs in that there is an improvement in the human animal bond. In addition, engaging in enrichment for our pets provides a source of distraction and entertainment for us as caretakers.[1] You cannot have a healthy animal if you are ignoring their mental health and their need to express normal behaviors. Enrichment is one easy way to ensure you are doing your best to meet your patient's mental needs, in addition to providing veterinary care.

What is Enrichment?

There are many definitions of enrichment depending on your source, but in simple terms enrichment can be defined as: to enhance the quality or value of something. In the case of our exotic companion animals, it means to enhance the quality of care by providing animate, inanimate, and nutritional environmental modifications for their well-being.[2] The focus of enrichment is to enhance an animal's mental health, which in the long-term will have an impact on their physical health as well. The goal is to provide a beneficial environment that will encourage our pets to express species-appropriate behaviors and mental activities, as well as to reduce abnormal behaviors and increase their ability to deal with changes or stressful situations.[1,2] It is important to express to an owner that enrichment is not an optional item, but rather, a necessary aspect of captivity.[3]

Benefits of Enrichment

Research has shown that there are multiple benefits of providing enrichment to not just our exotic companion animals, but all animals. Some of those benefits include the following:

- Increased behavioral diversity
- Decreased abnormal behaviors
- Decreased anxiety
- Improved learning abilities
- Reduced depression
- Mental stimulation
- Increased exercise
- Increased bonding to caretakers

Even small changes such as scattering or hiding food can have a positive impact on the behavioral expression of our exotic animal companions. Not only does enrichment enhance an animal's quality of life, but it is also a tool an owner can use to monitor the health of their animal. An animal that has historically enjoyed engaging in enrichment opportunities that is suddenly not interested in interacting with any of the provided stimuli might be indicating that it is not feeling well or is in pain.[1-5] Providing enrichment in a hospital setting can therefore make an animal more comfortable during its short visit and even extended stay by being able to express normal behaviors. This can in turn help veterinary staff assess the physical and mental well-being of the animal.

Potential Shortfalls of Enrichment

While there are many positive impacts to our exotic animal companions, we must acknowledge that there can be some potential issues of implementing enrichment activities. It might take some trial and error to find the best type of enrichment that will suit a particular species and stimulate normal behaviors. It can also take some time to build up to a proper enrichment protocol, since not only are you trying out which items work best for a particular animal, but it may also take time and money to create and obtain the different items.

Cost can be another concern as commercially made items can come with a hefty price tag. While obtaining items from pet stores or online are options, you can always recycle items you find at home or the veterinary clinic and create your own unique items that are tailored to the specific pet's preferences and even save some money in the process.[2]

In a clinic setting you will be limited by the time and material available, as well as the health status of your patient. It is always beneficial to have a conversation with the owners ahead of time to find out what kind of environmental enrichment their pet prefers.

Despite exercising caution, accidents can happen with new items, and animals can get injured with any of the newly offered enrichment items. It is important to always monitor an animal when you are offering a new item in their enclosure to ensure they are not going to get injured. Always make sure that items do not have small holes where a pet can get stuck and that the materials used are not toxic.[2]

One last important thing to remember is that while enrichment is aiming at reducing stress in the long term, it is not possible to eliminate all fear; it is, however, essential to reduce conditions of fear.[6] You need to be able to recognize the signs of a stressed animal as in the beginning stages some animals can be stressed with changes to their environment and might require a slower approach.

Types of Enrichment

Types of enrichment can be divided into two broad categories: social and non-social enrichment.[1] Social enrichment refers to either conspecific interactions or interactions with owners or other animals. Prior to introducing a new animal, it is important to consider the species' social preferences in the wild. Many different species such as rabbits, rats and chinchillas are very social and enjoy living with conspecifics, but individual preferences should be observed. On the other hand, species such as hamsters are not social and problems can arise when these animals are kept in groups. If keeping animals together is not an option, it can be beneficial to have animals close together but in separate cages. Animals can then safely interact at a distance. This means that any animals that are bonded will benefit from being at the veterinary clinic together if feasible.

Interactions with the owner or other caretakers (or the veterinary staff) need to be gentle and can include time spent petting, holding, or playing with the pet. These types of interactions are based on trust and need to be built slowly. Social interaction with the owner is important to reduce an animal's anxiety when around humans.[1,2] Providing positive reinforcement during those interactions will have a greater impact, even if you only have the ability to spend a couple of minutes interacting with the pet.

Non-social enrichment refers to the environment and all items a pet is exposed to within its enclosures or general living space; it includes things such as toys, beds, substrates, noise, and light.

General Enrichment Concepts

Any enrichment provided should take into consideration the natural history of the species and the range of normal behaviors that they exhibit. For most of our small exotic mammals we have a general idea of their wild counterparts and can extrapolate into care at home. While these guidelines can be implemented in a hospital setting, space, time, and medical conditions can limit the extent of environment enrichment.

Figure 1 showcases some enrichment options.

Figure1. Examples of environmental enrichment.

Enclosure

While there are resources outlining the minimum size of cage an owner should provide to their pet, they should consider obtaining the largest sized enclosure they are able to afford and have room for in their home. This allows for greater movement and even the ability to assign specific areas for particular functions.[3] Consider adding vertical space in the enclosure when possible and if appropriate for your particular species. This can be done by including ropes, hammocks, and extra floors.

The substrate you use will depend on the species you have. Many species like to dig so providing a deep enough substrate to allow for this behavior is recommended. Research has shown that animals prefer hard surfaces over wire surfaces, plus wire flooring is not typically recommended as it can cause pain and wounds on the pet's feet.[2,3] In the hospital we are going to be limited by the size of the already existing cages, but as it is for a short stay it will not be a problem.

Bedding

The amount of bedding provided can be enriching for animals such as rodents by providing a deep enough layer to allow for digging. One research study showed a preference for about 10 cm (4 inches) of soft bedding. Note that having a deeper layer does seem to have a greater impact since it can come into contact with the water bottle and get wet, potentially leading to hypothermia.

The ideal bedding material is dust free, non-palatable, absorbent, and free of toxic components. Different species and different medical conditions might require you to change the type of bedding that you use.[2] Again, for any animal in hospital we want to provide bedding that we can clean up easily and avoid potential cross contamination. Offering towels or soft fleece blankets are a great option.

Toys, huts, boxes, tubes

Providing different toys with different functions and different materials can encourage play in the enclosure. Ensure no metal or glass toys are provided as they could injure the animal.[3] Many rodents like to gnaw on things so providing safe sticks or wooden blocks to encourage this behavior is recommended. Rabbits like to toss toys around so providing stacking cups or other plastic toys can keep them entertained.

Huts, boxes and tubes of different sizes and materials are also enjoyed by many species to hide in during the day, when they are resting or if they are stressed. You also want to make sure you can easily clean the huts provided.[2,3] Ferrets particularly enjoy running through tubes so they should always have access to this enrichment.[3]

Exercise

While having species-appropriate items in the cage can encourage exercise, placing animals in a separate pen for exercise can further stimulate this activity by encouraging exploration. The exercise pen can be kept indoors or even outside, weather and safety permitting.

Rodents also enjoy running on an exercise wheel, so providing one can be a form of enrichment for them. Ensure that they are cleaned often and are of an appropriate size and material.[3]

Certain species such as ferrets, and rabbits can be trained to wear a harness and be taken out into the backyard for extra exercise while still being kept safe. Always monitor your pet while outside to ensure it does not get injured, overheat, or manage to escape.[3]

Larger exotic mammals that are hospitalized should be allowed to spend time out of their small cages as long as medically allowed and you can provide a safe area where they cannot get injured or escape from.

Light

Some exotic animal companions such as rodents and rabbits are usually more active during dusk and dawn, so they benefit from decreased light levels to encourage activity.[2]

Puzzle feeders and foraging toys

When we look at the natural history of the wild counterparts of our exotic animal companions, we find that they spend most of their waking time searching and foraging for food. We can encourage this behavior easily by spreading the food out around the enclosure instead of placing it all in a bowl. This encourages animals to move around and explore to find food items.[7] You can further challenge your pet by making more complicated toys. Simple strategies for this are using paper bags or egg cartons to hide food in or even using an empty toilet paper roll and filling it with hay and treats is a great alternative. There is also the option of acquiring commercially made puzzle feeders and adapting them to your pet. The ball treat dispensers are great for our larger exotic pet animal companions to distribute their pellets.

Training

Training is another great strategy to provide enrichment by stimulating your pet and increases the owner-pet bond. Find a high reward treat for your pet and start slowly by conducting short training sessions and slowly creating the behavior you want. This can be a fun experience and you can also train your pet to be comfortable with procedures such as regular nail trims and accepting medication via syringe.[8] Training sessions do not have to be long; they can be as short as 5-10 minutes and can easily be done during the pet's visit to the hospital.

Species-Specific Enrichment Suggestions

Each species will require different tactics to provide enrichment in their lives and Table 1 shows-examples of easy items to incorporate but it is not meant to be an all-inclusive list.[1-9] Please remember to always supervise your pet when you provide a new enrichment item. You want to ensure it does not stress them out and that all parts are safe.

Table 1. Enrichment examples for a variety of species.

Species	Enrichment examples
Ferret	Multi-level cage connected with ladders and ramps
	Appropriately sized tubes to run in
	Boxes with shredded paper
Rabbit	Supervised yard time for exercise
	Plastic toddler toys
	Box filled with shredded paper and hay for digging
	Chew sticks
Guinea pig	Hiding food in bedding
	Multiple homes to hide in
	Brushing
Chinchilla	Use of rope perches in the cage
	Dust baths
	Compressed hay cubes
Rodent	Tube maze within cage
	Objects to chew

Conclusions

As this paper discusses, environmental enrichment is a necessity for all our exotic animal companions to ensure they can express their normal behaviors and are able to cope with any changes in their lives without too much added stress. Many different options are available, and it is important to work with your selected species to be able to encourage them to express their normal behaviors.

References

1. Banks RE, Sharp JM, Doss SD, et al. *Exotic Small Mammal Care and Husbandry.* Ames, IA: Wiley-Blackwell; 2010:11-19.

2. Brandao J, Mayer J. Behaviour of rodents with an emphasis on enrichment. *J Exot Pet Med.* 2011;20(4): 256-269.

3. Pilny AA. Small exotic companion mammal wellness management and environmental enrichment. *Vet Clin North Am Exot Anim Pract.* 2015;18:245-254.

4. Lidfors L. Behavioural effects of environmental enrichment for individually caged rabbits. *Appl Anim Behav Sci.* 1997;52:157-169.

5. Oxbow Animal Health. All About Enrichment. Available at: https://www.oxbowanimalhealth.com/blog/all-about-enrichment-handout. Accessed March 15, 2021.

6. Morris CL, Grandin T, Irlbeck NA. Companion Animal Symposium: environmental enrichment for companion, exotic, and laboratory animals. *J Anim Sci.* 2011;89:4227-4238.

7. Rommers JM, Bracke MBM, Reuvekamp B, et al. Cage enrichment: rabbit does prefer straw or a compressed wood block. *World Rabbit Sci.* 2014;22:301-309.

8. Harris LM. Ferret wellness management and environmental enrichment. *Vet Clin North Am Exot Anim Pract.* 2015;18:233-244.

9. Belz EE, Jamilyn S, Kennell R, et al. Environmental enrichment lowers stress-responsive hormones in singly housed male and female rats. *Pharmacol Biochem Behav.* 2003;76:481-486.

AUTHOR INDEX